HEALING
WATER

FACTS ABOUT IONIZED WATER

I0490997

SUDESH MALIK

INDIA · SINGAPORE · MALAYSIA

Notion Press

No.8, 3rd Cross Street,
CIT Colony, Mylapore,
Chennai, Tamil Nadu – 600004

First Published by Notion Press 2020
Copyright © Sudesh Malik 2020
All Rights Reserved.

ISBN 978-1-64892-602-0

Disclaimer

Health books require medical disclaimers. While I may not have medical qualifications or formal training in the health field, what does it take to understand the journey to achieving great health? Knowledge is not limited or defined by a doctorate or a few letters behind one's name. I have learned everything that I know about health and wellness by extensive research, trial and error, and seemingly countless conversations with others about health. Where I fall short in formal training, I more than make up for it in my learnings through on-ground experiences. **Everything you read in this book are simply my observations from my own experiences about why Ionized Water is so healthy.**

Now, for the medical and legal disclaimer:

The purpose of this book is to educate. It is with the understanding that the publisher and the author shall have neither liability nor responsibility for any injury caused or alleged to be caused by information contained in this book. **This book is not intended in any way to serve as a replacement for professional medical advice.** Rather, it is meant to demonstrate that aging can be slowed and even reversed and that Great Health is achieved when the most fundamental nutritional needs of the human body are met. If you feel the need, always consult a doctor or another medical professional when you have an illness or disease of any kind. We admit to knowing little if anything about medicine and therefore would never offer medical advice to anyone for any reason. Medicine is in one direction, health is in the other. **The author offers health advice that is his personal opinion.**

Dedicated to my mother Rajshree

and my daughters

Anupa Vohra

Neha Kadakia

Drishti Malik

Nishka Kapur

Kaisha Irani

Deesha Malik

Contents

Thank You

This book and the work towards ionized Water would be incomplete without the dedication and support of my Miracle Team Members from all over India. Your passion for this mission has brought you all from different walks of life to volunteer your time, energy and effort so that many others may reap the benefits of ionized water. I extend my deepest and sincerest gratitude to each and every one of you for all that you do. Thank you for being a part of this journey. I look forward to journeying even further with you.

My special thanks to

Shailendra Tiwari	Indore	Tarun Adwani	Mumbai
Satpal Sagar	Delhi	Deepak Singh	Patna
Dr.Acharya	Kolkata	Y.K Pradhan	Orissa
Daulat Tungaria	Jaipur	Santosh Mahanti	Chhattisgarh
Akant Pathak	Bhopal	Prashant Sah Banka	Bhagalpur
Dnyaneshwar Dorkar	Nagpur	Sanjay Nangyan	Roorkee
Deepak & Kanchan Bambani	Mumbai	Arun Gupta	Delhi
Mahesh Hada	Indore	PremPal Singh	Agra
Dushyant Sahu	Indore	Arun Kumar	Bhubneshwar
Ranveer Singh	Indore	Meghana Sheth	Gujarat

Thank you once again for being a part of our mission which is spreading good health in the most easy and natural way I.e. drinking water.

About us

Miracle is a group of independent group of like minded people across India. Our mission is to help families get access to better health through Ionized Water. Founded by Mr. Sudesh Malik, we are a Mumbai-based team with active full-time and part time volunteers across the country. Our members consist of esteemed people from various professions including but not limited to – scientists, engineers, businessmen, health enthusiasts, teachers and homemakers. Through the efforts of Miracle Team, families across the country today are enjoying good health and well-being.

Our team conducts free wellness seminars to educate the masses on how easy it is to stay healthy. We also conduct regular programs to train people to spread this message effectively. Miracle team regularly produces a wealth of educational material across its Facebook and YouTube channels to spread the message of the importance of good health and to bring about a shift in lifestyle for the better.

Introduction

Gone are the days when plain tap water was enough of an elixir to nourish the body and replenish the system with nutrients. As a millennial generation advances in years, we're starting to understand the importance of extracting every ounce of nutrition from even their drinking water.

Simultaneously, all are waking up to the importance of prioritizing their health, in the face of a fast-paced lifestyle filled with stress, pollution, questionable food habits and unhealthy work hours. After all, if you're supposed to drink 2 to 4 liters of water per day as per your weight, wouldn't you want it to give you a boost of nutrition, vitality, improve your health and keep diseases at bay?

There are toxins in everything we consume today – our foods, beverages and even acidic water. These toxins work slowly, over time to weaken our immune system and make the body prone to illnesses. We're made of 70% water, we're constantly absorbing things from our environment, what we eat, what we breathe and so on. The average city water supply contains parasites, viruses, pesticides, hormones and toxins that aren't entirely killed by chlorination alone. It's up to us to go a step further in improving the quality of water we consume, and therefore improving the quality of life.

In Japan alone hospitals and clinics have used electrically restructured, hexagonal, alkaline water successfully for over 40 years to combat various diseases.

Water is the basis of all bodily fluids and functions, including blood, lymph and digestive fluids. It's essential for the transportation and absorption of nutrients as well as elimination of waste. Water keeps our body temperature regulated, joints lubricated, organs cushioned and tissues moistened. It helps carry and transmit signals throughout our body. The movement of water in and out of cells produces significant amount of energy, which is why fatigue is one of the first and most significant signs of dehydration.

Ionized alkaline and acidic water is not only safer for consumption, but also allows our bodies to reach homeostasis i.e. the perfect balance with our environment, better. Conventional medicine, physiologists and natural health experts suggest that even a subtle shift in the Ph balance of the body can affect overall health and well-being, resulting in chronic fatigue, pain, weight fluctuations and decline of athletic performance. The water you drink every single day for all your life plays a crucial role in maintaining optimal Ph levels in your body on a long-term basis.

It's time to rethink the way you drink your water!

Life in a Fish Bowl as explained by Dr. Dave Carpenter

 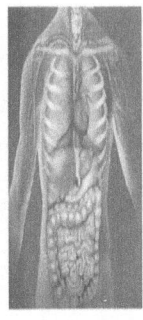 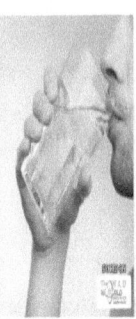

Think of your environment as a fish bowl. If the water is not changed or filtered regularly, wastes build up and can eventually extinguish the life within. One of the first functions that is put on "hold" when water is lacking is detoxification – the ongoing removal of cellular waste and environmental toxins.

All the detoxification pathways in the body (liver-colon, kidney-bladder, skin-sweat, lung-breath, and lymphatic system) require water. When water is not supplied in abundance, wastes build up in the fluid that surrounds each cell and all the detoxification pathways become sluggish. However, the body is ingenious. It always adapts. Under the stress of dehydration your body will find places to store toxins where they will not immediately interfere with critical life processes. Toxins and wastes can end up in fatty tissue, in joints, and as deposits in arteries. In the short term, life is preserved, but the long-term consequences are obvious.

Drinking plenty of good water every day is like changing the water in your fish bowl. And as long as you need to drink water, you might as well get the best water possible. For a variety of reasons, ionized water is being recognized by professionals as the best choice.

Taking Responsibility For The Fish Bowl

Everyone has seen a fish bowl that needs cleaning. Slime builds up on the glass and wastes cloud the water. The longer you wait to clean the bowl, the more difficult the task becomes, and the greater the chances of losing the fish. In this analogy, the fish is you, and the cleanliness of the fish bowl is an indicator of your health. Ultimately, you are the only one who can take responsibility for keeping the fish bowl clean and for keeping the fish alive.

Water is the foundation of every health care program. Since you are 75% water, drinking the best water you can find is a big part of keeping your fish bowl clean. Drinking restructured, ionized alkaline water is the easiest single thing you can do to support the environment in your fish bowl.

Chapter 1

What is a Ionizer

The Root Cause of Chronic Disease

Stress is defined as a straining force, an imbalance of forces.

The root cause of disease is unchecked stress.

Stress is a form of opposition that stimulates adaptation, growth and development in any life form exposed to it. Stress is only "bad" when it is excessive, unbalanced and unmitigated.

Stress is necessary – it helps us function and get things done. Think of it this way – the stress of pressure turns graphite into diamonds. The stress of sunlight is captured by chlorophyll in plants and converted into an energy form (glucose) that feeds the growth of the plant. The stress of the birthing process gives rise to new life. It was stress that formed the Grand Canyon, Niagara Falls and the Hawaiian Islands. Unless stress is applied onto muscles, it doesn't grow and develop.

Stress is a necessary stimulus for life. It is a catalyst for beauty and growth. However, when stress exceeds our limits to contain and manage it, disharmony and imbalance result. Three of the most common types of stress that I have seen and managed, namely:

1. Oxidative Stress

2. Acidic Stress

3. Hydration Stress

In order to learn to manage one's stress effectively, it's important to first understand each form of stress. The more stress accumulates in our systems, the more our cells are distressed and the more effort is required for our body to achieve its balance so that it may heal well.

Alkaline Ionized Water is by far the most nutritionally rich drinking water that is available. Ionized water is the answer to all the above stresses made through a process called electrolysis.

Let us understand electrolysis

Tap water, as we know it, contains two kinds of ions: mineral ions (calcium, magnesium and potassium) and bicarbonate ions. The former carries positive electromagnetic charge, while the latter carries a negative electromagnetic charge. Domestic water ionization systems use electromagnetism to separate positively charged minerals from the negatively charged bicarbonate. Water ionizers are fitted with charged plates that perform the role of magnets. They attract the opposite charge; when this happens, ions pass through a membrane of what we call Bipolar Exchange Membrane, which only allows the passage of ions and filters out other type of particles.

This process creates alkaline and acidic water simultaneously – by transferring each type into its respective chamber.

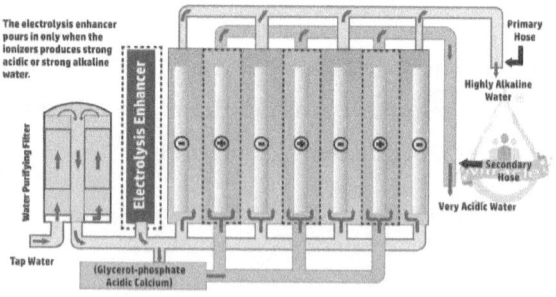

When producing strong acidic water, the primary hose will produce highly alkaline water while the secondary hose produces the very acidic water.

Titanium plates dipped in Platinum.

The process consist of regular water is runing over negative (cathode) and positive (anode) tubes, to give it a charge. The minerals in the water get ionized (acquire charge) in the process, creating positive ions (hydrogen) and negative ions (hydroxyl). The electrodes are made of titanium, said to be the hardest metal known, and then dipped in platinum which is an excellent conductor of electricity.

Essentially, what happens during electrolysis is that water molecules are restructured through the process of ionization. In most commercial processes, water is first run through a high-grade filter to remove traces of organic and inorganic contaminants and microbes ever-present in regular tap water. This filtered water is then run over a series of platinum – coated titanium electrodes. If the water contains minerals like calcium, magnesium and sodium, electric current can be passed through the water and the process of electrolysis is completed. If the water does not contain minerals, then no electric current can pass through it. Such water is called 'dead water'.

Electrolysis uses high amounts of energy and a catalyst to change the molecular structure of water. It does so in the following ways:

1. Ph: Electric current splits water into Hydrogen (H+) ions and Hydroxyl (OH-) ions

2. ORP: Platinum nanoparticles from the electrodes generate free electrons which associate with hydrogen ions to form 'Active Hydrogen'

3. Micro-clustering: The magnetic resonance energy of water molecules is reduced

As mentioned earlier, the reaction between the cathode and water molecules gives rise to Active Hydrogen and Hydroxyl Ions, resulting in an electron-donating alkaline solution. Meanwhile, the anode reacts with water molecules and gives rise to Active Oxygen and Hydrogen Ions resulting in an oxidized, acidic solution. Basically, through this process of electrolysis, we get two different types of water. Under ideal circumstances, the combined pH of these two water types should come to 14.

For example, if one hose generates water, with an alkaline pH of 8.5, the other hose will deliver water with a pH of 5.5.

So, what's the difference between Alkaline Water and Acidic Water? Alkaline water is rich in antioxidants, thanks to the active hydrogen. This water is ideal for consumption, as it supports and enables cellular healing by alkalizing, hydrating and neutralising free radical damage. Acidic water on the other hand, is rich in oxidizing agents thanks to the active oxygen.

This water carries bactericidal properties due to its low pH and oxidizing effect on infectious microbes, making it ideal for topical use, such as treating wounds, cleansing skin etc.

What is Ionized Water?

As we've understood earlier, ionization means to gain or lose an electron and acquire an electric charge. Depending on the charge it carries, ionized water can be classified as Alkaline and Acidic water. Both possess extraordinary benefits – however, both serve very different and unique purposes.

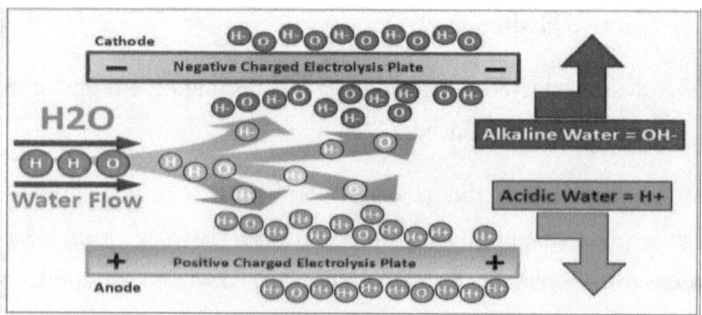

Alkaline ionized water is used for internal healing, boosting and repairing the body and is ideal for daily, regular consumption. Acidic water, on the other hand, should be reserved for external use only; such as acne, scrapes and wounds, rashes, etc. It's anti-bacterial in nature and can be used for gardening as well, since it encourages plant growth.

Naturally Ionized

Ionized Water is not a manmade concept or a synthetic concoction. Ionized Water is found in abundance in nature. When water bounces off rocks and tumbles down mountains, it picks up electrons from the earth which is already our greatest source of free electrons. The speedy movement of water against earth gives water an abundance of negative ions, which has a powerful effect on the health and well-being of those who consume it. While it's unfortunate that most of us don't live by a waterfall, a high-quality water ionizer at home is the next best thing. It purifies the water in your home, and turns your tap into a negative ion generator.

This water helps you flush out acidic waste from your body to neutralize free radicals and structure water for far superior hydration, enhanced nutrient absorption, more effective detoxification, increased metabolic efficiency, and improved cellular communication. It literally turns your faucet into a fountain of youth.

You may come across Ionized Water under many other names, such as:

Electrolyzed Water, Reduced Water, Ionic water, Hydroxyl Water, Alkaline Water, Alkali Water, Alkalized Water, Cluster Water, Micro cluster Water, Micro Water, Ion Water, Electron Water and etc.

Ionized Water and Our Bodies

When we drink sufficient amounts of ionized water, we start flushing our body's cells with alkalinity and antioxidants, simultaneously hydrating our system too. This seamless process of cleansing and replenishing keeps our body healthy and running smoothly.

Make no mistake, Ionized Water is not a manmade concept or a synthetic creation. It is available abundantly in nature in the mountain waterfalls, streams and springs. How? Water bounces off rocks, resulting in mild ionization and the creation of hydroxyl ions. This is what makes water from these sources so enriching for our bodies.

Ionized Water has three main properties:

1. Powerful Antioxidant

2. High pH (Alkaline)

3. Nano molecules or Micro-clustered

1. Oxidation-Reduction Potential (ORP)

Your body's cells face threats every day. Viruses and infections attack them. Free radicals also can damage your cells and DNA. Some cells can heal from the damage, while others cannot. Scientists believe molecules called free radicals can contribute to the aging process. They also may play a part in diseases, like cancer, diabetes, and heart disease.

Antioxidants are chemicals that help stop or limit damage caused by free radicals. Your body uses antioxidants to balance free radicals. This keeps them from causing damage to other cells. Antioxidants can protect and reverse some of the damage. They also boost your immunity.

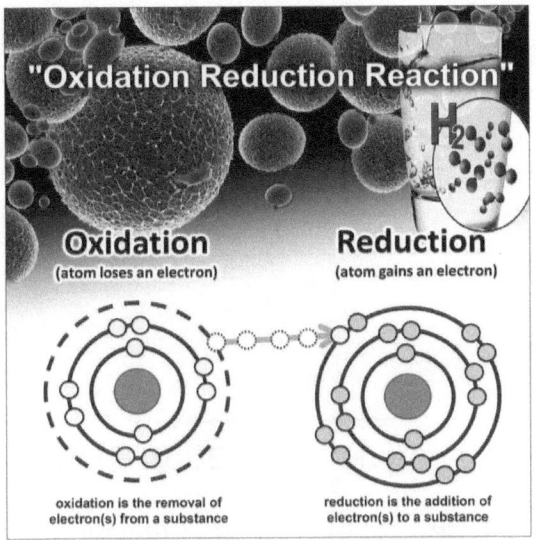

The unit to measure oxidation in water is ORP (Oxidation Reduction Potential) It is a measurement of the strength of the oxidizing or reducing power of a solution. Oxidation and reduction reactions involve an exchange of electrons. A negative ORP measurement indicates a surplus of electrons the more negative the number is, the greater the number of surplus electrons. A positive ORP measurement indicates instability. A solution with a positive ORP is hungry for electrons.

Water that's rich in active hydrogen thus carries negative ORP. What does this mean for us? The more free-radical neutralising electrons water contains, the more potent with antioxidants it becomes. This water has the ability to react with other compounds in the body and control redox reactions. Essentially, what these electrons do is that they create an electric field inside a cell, acting as a rich source of antioxidants and neutralising some of the harmful effects of free radicals. By controlling free radicals itself, ionized water increases the lifespan of cells and strengthens our body.

OXIDATION REDUCTION POTENTIAL

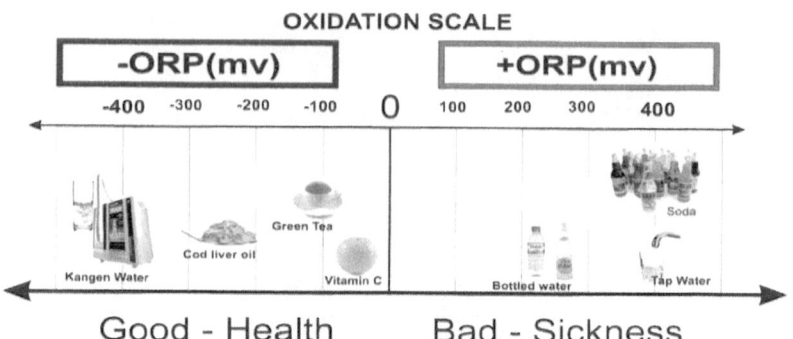

Good - Health   Bad - Sickness

In stark contrast, most bottled water, sports drinks and soft drinks carry highly positive ORP values. Unlike water that carries negative ORP, positive ORP-carrying fluids create free radicals and add to oxidative stress. This means it has the reverse effect on your body, as compared to negative ORP-carrying water.

Oxidation and reduction are phenomena that we actually witness all the time. Oxidation breaks down substances. For example: rusting of iron and apples turning brown when cut. When unstable atoms that lack electrons take electrons from other atoms to achieve stability, oxidation occurs. This process keeps going until every atom has achieved stability. To put it in context of say, a rusting iron. The chain reaction of oxidation will go on and on until the iron has completely rusted.

A similar process is taking place within our bodies. This oxidation is what caused aging. But, oxidation is also a necessary function of the immune system which strips electrons from bacteria and other invaders to halt the growth of infections inside of our bodies. The opposite of oxidation is reduction. Reduction occurs when bacteria have been destroyed. Antioxidants step in to donate electrons and put a plug on the chain reaction. If your body has insufficient antioxidants, the oxidation goes on, thus destroying healthy tissue. The destruction of healthy tissue is known as free radical damage. This explains infections and the process of aging.

The core function of antioxidants is to supply electrons to electron-deficient free radicals so they no longer strip electrons from vital cells. The best antioxidants are ones that are easily and abundantly bio-available and have higher electron-donation capacity. Antioxidant supplements are a multi-million-dollar industry globally.

Besides being a measure of electron activity, ORP also helps us to measure the antioxidant capacity of a solution. Negative ORP values indicate abundance of available electrons, and any solution with a Negative ORP is an antioxidant.

Most bottled waters have an ORP between +150 and +300 millivolts (mV). Tap water can have an ORP value as high as +500. What does this mean? To put it simply, with positive ORP values so high, you're basically getting free radicals in a bottle. On the other hand, freshly squeezed orange juice has an ORP between – 200 and – 100. And we all know the antioxidant goodness of orange and similar citrus fruits. However, when a citrus juice is left exposed to the air, it begins to oxidize and lose its antioxidant power. Processed orange juice has an ORP of about +200 mV – far from being an antioxidant.

When alkaline water undergoes ionization, it gains extra electrons and becomes a very potent antioxidant. This, restructured, ionized water has an ORP between – 300 and – 800 depending on the original composition of water. What does this mean? A glass of alkaline ionized water has much more antioxidant capacity than a glass of freshly squeezed orange juice. Regular consumption of this water can help fight inflammation, pain, disease, infections and other symptoms and causes of free radical damage. You no longer have to stock up on exotic ingredients and supplements, all you need to do is drink water!

There's one more very crucial benefit of drinking Negative-ORP Water – intestinal health. A surprisingly large number of digestive problems are due to a disturbance in the gut fauna in the digestive tract. The easiest fix to most of these problems is to replenish and maintain a healthy environment for these bacteria to flourish. How can we achieve this? 95% of gut bacteria are anaerobic, which means they need negative ORP values to flourish. Consuming Negative-ORP Water and foods is therefore the easiest and most effective solution to maintain gut health.

What about acidic water? While alkaline water gains electrons acidic water loses it. This creates a solution with a Positive ORP value. Your water goes from antioxidant to free radical. And we obviously don't want that. Having said that, acidic water is not without its own fair share of purpose. Strong acidic water is a proven antibacterial, anti-microbial, and anti-fungal for use on skin, foods, plants, counter tops, etc. It has been used as a disinfectant in many hospitals and restaurants for decades and was featured in the February 23, 2009 edition of the LA Times. Besides being a great antibacterial wash (one of the best there is because it contains no toxic substances or antibiotics), strong acidic water is great to have around the kitchen. It works well to disinfect counter tops and cutting boards. It is also a wonderful aid to cleaning. The natural acidity of the water helps to remove hard water deposits on glass and other surfaces.

2. Alkaline pH (Potential Hydrogen)

According to the chemistry of electrolysis, positively charged hydrogen ions are attracted to the anode, whereas negatively charged hydroxyl ions are attracted to the cathode. Thus, electrolysis separates water into acidic (positively charged) and alkaline (negatively charged).

pH 7.0 is neutral, equal (H+) & (OH-) ions pH 2.5 is very acidic, many (H+) ions

pH 11.5 is very basic, many (OH-) ions

The term pH stands for "potential Hydrogen." The more hydrogen (H+) ions are concentrated in a solution, the more acidic it is. While the full pH scale ranges from 0 to 14, anything below a pH of seven is considered acidic and anything above seven is considered alkaline (or basic). Therefore, a solution with a pH of 2.5 is much more acidic and has much more hydrogen (H+) ions than a solution of 6.5. Likewise, a pH of 11.5 has more hydroxyl (OH-) ions and is therefore much more alkaline than a pH of 7.5.

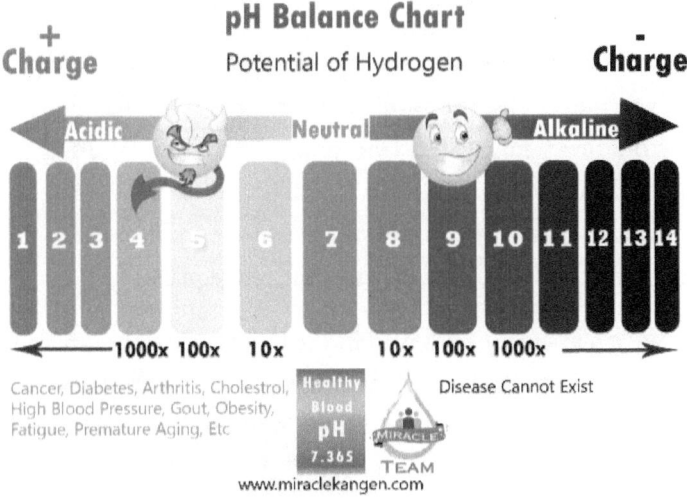

There is a logarithmic relationship with reference to pH scale. For example, for every one-point difference in pH there is a HO-fold difference in the number of ions present. Put another way:

pH 6 has 10 times more hydrogen *(H+)* ions than a pH of 7 pH 5 has 100 times more hydrogen *(H+)* ions than a pH of 7 pH 4 has 1,000 times more hydrogen *(H+)* ions than a pH of 7 pH 3 Has 10,000 times more hydrogen *(H+)* ions than a pH of 7 pH 2 has 100,000 times more hydrogen *(H+)* ions than a pH of 7 pH 1 has 1,000,000 times more hydrogen *(H+)* ions than a pH of 7

Similarly,

pH 8 has 10 times more hydroxyl (OH-) ions than a pH of 7
pH 9 has 100 times more hydroxyl (OH-) ions than a pH of 7
pH 10 has 1,000 times more hydroxyl (OH-) ions than a pH of 7
pH 11 has 10,000 times more hydroxyl (OH-) ions than a pH of 7
pH 12 has 100,000 times more hydroxyl (OH-) ions than a pH of7
pH 13 has 1,000,000 times more hydroxyl (OH-) ions than a pH of 7

Therefore, by nature both acid and alkaline water can be created as electrolysis moves electrons and separates acid from alkaline on each side of the membrane. Since the total pH must total 14, whatever alkaline pH number is runs out of the alkaline hose, the water coming from the acidic hose, by convention, must have a pH of the difference, so that when combined with the alkaline pH, equals 14.

pH balance plays a crucial health in the body's overall health and well-being. The normal pH range of a healthy human is between 7.30 and 7.45. The pH level of your blood indicates how much oxygen is available in your blood cells. At a pH of 7.45, your blood contains 65% more oxygen than blood at a pH of 7.34.

What does this tell us?

Lack of oxygen results in metabolic acidosis, which is a component of just about every disease we know.

Blood that has a pH of 7.3 is also thinner than blood with a more acidic pH. Acid is what is partially responsible for blood clothing. The lower the blood pH, the thicker the blood. And thicker blood does not flow easily, and is more difficult to pump, which in turn puts extra strain on your heart, leading to hypertension and high blood pressure. Dehydration and low blood pH are some of the greatest contributors to high blood pressure.

Given our current lifestyle and habits, most people's bodies and interstitial fluids are higher on the acidic side than can be considered healthy. This is a direct consequence of a diet that's heavy in processed and refined sugar, aerated drinks, excess proteins and refined carbohydrates, along with unmanaged stress, dehydration and exposure to pollutants. We can see the side effects of this lifestyle every day – premature aging, lower stamina, sluggishness, fatigue, high susceptibility to illnesses. The body also slows down in its ability to absorb nutrients and minerals, produce energy and regenerate cells. Acidosis hardens arteries and makes you easily prone to fatigue and weakness. When the body is overly acidic, it becomes incredibly difficult to maintain normal blood pH. The body has no choice but to pull alkaline minerals from organs, tissues, bones and teeth, causing your body to basically slowly disintegrate.

How can Ionized Water help? When water is ionized, positively charged alkaline minerals are attracted to negatively charged hydroxyl ions. Alkaline water, thus, has a pH of 7 and is enough to replenish the body's mineral levels and neutralize acidic waste in your body. Alkaline Water contains more oxygen. At pH 7, it has equal number of hydrogen and hydroxyl ions. As the pH increases, the hydroxyl ions increase. When you drink alkaline water, you are drinking water with more oxygen – not in the form of oxygen but in the form of hydroxyl, which is stabilized due to its combination with an alkaline mineral.

Once ingested, these hydroxyl ions can form a water molecule and release an oxygen atom in the process. Thus, alkaline water can be used to neutralize acids and supply oxygen to the cells.

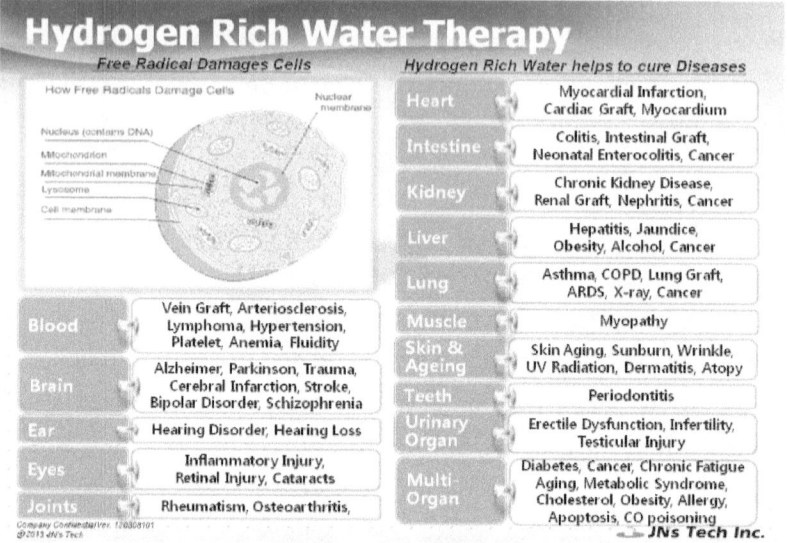

Can Ionized Water Neutralize the Acid in Your Stomach?

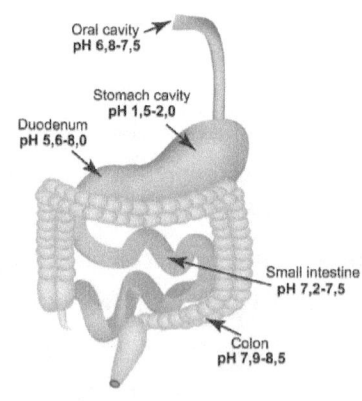

The answer is simple – the cells in your stomach produce concentrated Hydrochloric Acid pretty much on a continuous basis or whenever it has to digest food. When you eat or drink, the pH inside your stomach becomes alkaline. The increase in pH levels along with the expansion of your stomach stimulates the secretion of HCL, in order to maintain pH levels. When alkaline water is consumed, the stomach stimulates more HCL to be secreted. What's interesting to note is that this reaction gives us bicarbonates as a by-product. Carbon dioxide, water and salt are used up in this reaction, which in turn gives us HCL and sodium bicarbonate in the bloodstream.

Chemical reaction that produces stomach acid:				
NaCL	+H2O	+CO2	=HCL	+NaHCO3
(Salt/Sodium Chloride)	(Water)	(Carbon Dioxide)	(Hydrochloric acid)	(Sodium Bicarbonate)

With the progression of time and age, we lose the bicarbonate buffers, which is a sign of increasing acidity. In this case, our primary goal is to support the blood's buffering capacity, so it can neutralise acid levels and slow down the symptoms of disease and aging.

Anyone who consumes alkaline water benefits in many ways:

1. It provides essential minerals to neutralise acidic waste in the tissue and blood. This also takes the pressure off bones, organs, tissues and teeth.

2. Alkaline water can carry more oxygen to the organs in the form of – OH ions.

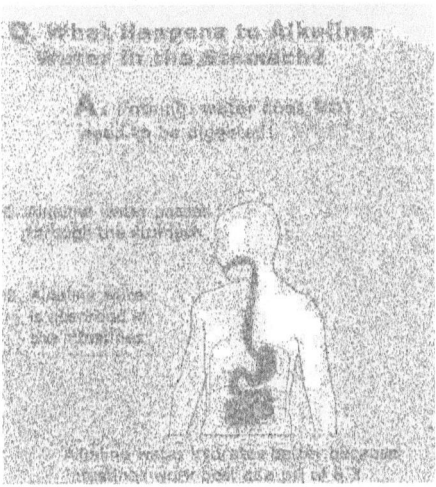

3. Consuming alkaline water (especially 20 minutes before a meal) stimulates the production of Hydrochloric Acid faster, which in turn helps in better digestion and assimilation of nutrients. Did you know that most North Americans over the age of 40 don't produce enough HCL in their stomachs?

3. Micro-Clustering

Perhaps one of the most significant features of ionized water is its ability to hydrate cells. The precise explanation for this mechanism has not been clearly established. Since we now understand that water penetrates cell membranes through specific protein channels, it is conceivable that in a state of oxidative stress these protein pores may become damaged and somewhat dysfunctional. This could theoretically impair cellular hydration.

When we consume antioxidant rich alkaline water, it is not a stretch to imagine that these oxidized protein channels could be restored to their healthy state, allowing for improved entry of water into the cell, thus promoting optimal cellular hydration.

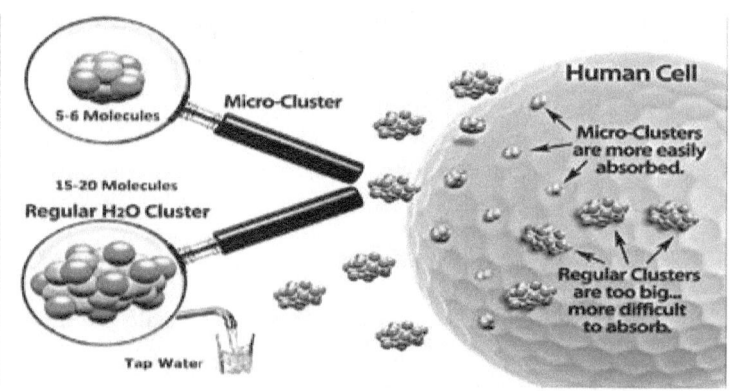

Another theory proposed by two different groups maintains that the electrolysis process lowers the vibratory frequency of the water molecule. This permits a closer association between molecules primarily forming a complex of six water molecules bound by its polar nature. This is referred to as *hexagonal water* or a *micro-cluster*.

Every molecule of water is made up of two hydrogen atoms and one oxygen atom to form the H2O molecule. The water we drink, comes in clusters of molecules, often as large as a cluster of 13 molecules. This is why you usually feel full for longer after drinking a glass of tap water, owing to the amount of water molecules needed to be processed for hydration. Due to the high volume of water molecules that the body needs to break down, the cells don't end up absorbing most of the water you drink.

On the other hand, micro-clustered water contains only five or six clusters of molecules. This water is absorbed easily, and is therefore much more hydrating, soluble and permeating for the cells and tissues. To put it simply, micro-clustered water is basically water with lesser volume of molecules but a higher rate of absorption. Micro-clustered water hydrates the body immediately, prevents bloating and enhances energy levels.

These water molecules are also high in antioxidants which can improve your overall aerobic capacity, boost your health, fortify your immune system and prevent premature aging.

Chemists call water a 'polar' molecule. This means can be both positively and negatively charged, because of the way the electrons are arranged around each atom. As a result of this, when water is placed in an electric field, the molecules arrange themselves in a very specific manner to stay balanced with the ionizing field. Water molecules form a crystalline, hexagonal pattern that can pretty much change its basic properties. However, even when the structure of water molecules dissipates after the electric field has been removed, water retains some of the newly acquired properties and structure. The crystalline structure improves water's ability to hydrate cells, transport nutrients, get rid of waste, support metabolic processes and improve inter-cellular communication.

Components of Ionized Alkaline Water

Alkaline dissolves acid in the blood with four main components, namely – Calcium, Magnesium, Potassium, and Sodium.

Calcium is needed for strong bones and teeth. It also important for heart health.

Magnesium converts food into energy and helps maintain muscle, heart, kidney and nerve functions. It is a mineral that is also important for normal bone structure in the body.

Sodium is both an electrolyte and a mineral. It controls blood pressure and blood volume. It helps keep the water and electrolyte balance of the body. Sodium also contributes to how nerves and muscles work.

Potassium is a very important mineral for the proper function, tissue, cells and organs in the human body. It also an electrolyte, a substance that conducts electricity in the body along with sodium chloride, calcium and magnesium.

The Difference between Alkaline Water and ionized Alkaline water and the scientific evidence behind it:

Alkaline water: Alkaline water is any water that has PH of 7.0 or greater. You can make your water alkaline by adding baking soda to it or alkaline minerals to your water.

Alkaline Ionized Water: Alkaline ionized water is water that has been filtered of toxins and chemicals and goes through a process known as "electrolysis in a water ionizer machine.

Alkaline water is made by adding chemicals and is NOT the same as Alkaline Ionized Water. There are little or no health benefits to drinking the former, except to the retailer selling it!

The key here is "Ionized Alkaline Water". "Alkaline water" and "Ionized Alkaline Water" are often used interchangeably and are easily confused by uneducated consumers. Ionized Alkaline Water and Alkaline water are distinctly different.

How Can Ionized Water Help You?

In Japanese, the word *Kangen* means 'return to origin'. Nothing comes closer to that promise better than restructured, ionized, alkaline water. It probably the closest we can get to the mythical fountain of youth. My own experience, and that of my patients is testimony to this. Ionized Water is restorative in nature. It supports every other modality in helping people return to their original health and vitality.

Energy

One of the earliest changes people notice when they consume ionized water is a spike in their energy levels. Even when the day is done, they are not. You can imagine what that must mean for people who are otherwise already drained and drooping by the time afternoon rolls in. The restructuring of water molecules is what makes all the difference. When the very essence, the very foundation of all bodily functions has improved, you are bound to feel better too. We don't give hydration enough credit for the role it plays in our health and vitality. Another interesting thing that most people who consume Ionized Water notice is that they also get a restful night's sleep. With improvement in both, energy levels and sleep, most of the stress is not so overbearing anymore and this alone begins to yield dramatically better results in your lifestyle.

Detoxification

Our bodies are naturally wired to be autonomous detoxifiers. However, our current lifestyle and the nutritionally-scarce foods and beverages we consume has put extra strain on our body's systems, which aren't able to function as efficiently as they should. This is why, in today's day and age, our body needs a bit of extra help with detoxifying.

Many scientists believe that environmental toxins and pollutants have given rise to new disorders, autoimmune diseases, mood disorders, chemical sensitivities and a whole new crop of syndromes. These conditions arise when the body's detoxification pathways are overburdened. A simple thing to mitigate this situation is to reduce the toxic burden. How do we do this? Good quality water is the most potent way to support the body's natural cleansing and elimination system. Water hydrates the blood cells and lymph so toxins can move rapidly through the detoxification pathways. It's also a major component of enzymatic processes where toxic compounds are broken down. Water is also crucial in the final stage of elimination, as it helps lubricate the intestinal walls and is the basis of excretion via urination. All of the above processes will naturally function better when the quality of water is better.

The body uses the same mechanism to flush out stored toxins that it uses to remove bacteria and viruses, so cleansing symptoms can often be similar to the cold or flu. This could mean headache, fatigue, skin breakouts, coughing spells, loose bowels etc. If these are symptoms you experience all too frequently, you should increase your intake of Ionized Water.

A Quicker 'Return to Origin'

Regular consumption of Ionized Water also helps in speedy recovery. And this is something I have noticed in my own experience with myself and other people. In some cases, six months of recovery is cut short to just six weeks. Their return of 'origin'al health is much rapider.

The deeper I dive into understanding restructured Ionized Water, the more I grow convinced that it will support recovery from most health difficulties and ailments. However, conditions linked with dehydration and/or acidosis show much faster recovery compared to other conditions. What are these conditions? Blood sugar problems, asthma and allergies, high blood pressure, skin problems, digestive and intestinal disorders, arthritis and other joint problems.

How Much Ionized Water Should You Drink? And When?

Healthy people should consume half their body weight of Ionized Water in ounces every day. For example, if you weigh 50 kgs, you should drink 2.5 liters of Ionized Water every day. If you're involved in physically active and gruelling work, or if you have health issues*, you should up your intake of Ionized Water.

When is the best time to drink Ionized Water, or any water for that matter? Ideally, you should be drinking this the first thing after you wake up. It will hydrate your digestive system before you eat breakfast. Drinking alkaline water 20 to 30 minutes before a meal can improve digestion vis a vis stimulation of hydrochloric acid for better assimilation of nutrients. It also gives your body adequate water for smooth digestion and to buffer the waste products cells excrete into the bloodstream. Avoid drinking water DURING a meal, but if you have to, go for water with neutral pH. Alkaline water will neutralize the acids required for digestion. Even neutral water will do this to a certain extent this is why drinking with meals is not recommended. Once you've consumed a meal, wait for another 30 minutes before consuming Ionized Water.

If you have health issues that require you to restrict your fluid intake, please follow your physician's advice. Just drink ionized, alkaline water for your fluid allowance.

Another equally crucial time for consuming Ionized Water is when you consume alcoholic or aerated drinks. Did you know that it takes 32 glasses of 9.5 pH water to neutralise ONE glass of soft drink? THAT's how acidic they are! (ref. Whang S. "Reverse aging 1990, pg. 55)

For long now, Alkaline Water has been referred to as the 'Hangover Cure' in Japan. When we drink Alkaline Water after consuming alcohol, it neutralises the excess acids while electrons neutralise the free radicals. Any impending hangover is therefore diminished or completely eliminated. One simple but effective way to incorporate Alkaline Water in your drinks is by freezing them as ice cubes. In addition to all the above recommendations, you should make it a point to consume water throughout the day. Keep a tumbler handy at your desk or wherever you spend most of your time.

Chapter 2

How to Choose Water Ionizer

For ionization watch out for 2 factors:

1. Plate

Ionization plates comes in 3 main styles – solid, mesh and slotted (also called hybrid).

2. Power

More power per square inch results are seen on pH and ORP output. Since electrolysis is the core of the process of ionization, the power these plates receive plays a very crucial role in the process. Even the best plates are of little use if the power is low. The more power the plates receive, the more efficient the machine is in separating the ions creating a high pH and negative ORP.

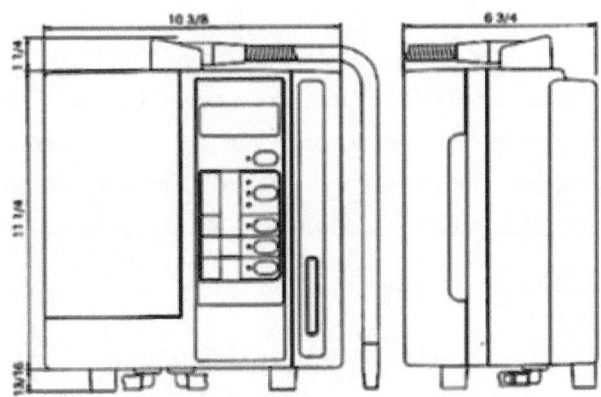

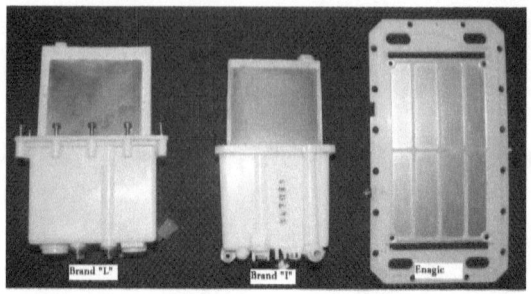

Types of Plates

Solid/Flat Plates:

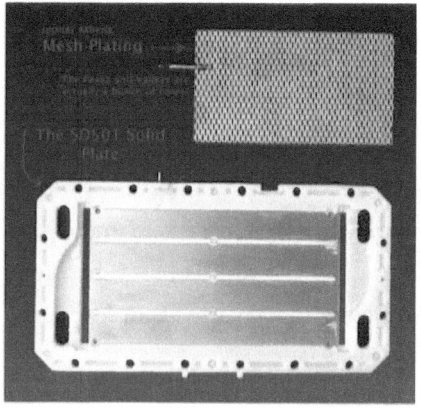

This is the first-generation technology. Originally, all water alkaline machines used large solid plates. It's known for its strength and durability, but the downside to these plates is that electricity focuses on the outer edges only, making it less efficient in generating ample antioxidants in the water, this could be easily overcomed by using good electric power.

Mesh/Slotted Plates:

Mesh plates technology is considered an advanced version of conventional plates design. Before the solid plates are electroplated, they are treated with mesh plating. Just like window screens, mesh plates have many edges that are excellent conductors of electricity. Unlike solid plates, the entirety of mesh plates conduct electricity making the process of electrolysis far more efficient and effective. Mesh design allows the electricity to pass through the whole area, the surface area enlarges, the electrolysis improves, the hindside is that the mesh plates are not strong and are more susceptible to wear and tear, thus exposing the titanium to water. Also mesh plates have a tendacy to get clogged and is difficult to clean, hence after a period of usage devices with mesh plates tend to function poorly.

Slotted/HYBRID Plates

Hybrid plates, as the name suggests, is a combination of both solid and mesh designs. The only difference is that the titanium plates contain electrolysis distribution holes. To create conductive edges, holes are drilled onto the plates. The holes do not affect the durability of the plates but it boosts the overall surface area.

They are large enough not to get clogged with limescale. After drilling, the plates are dipped in platinum several times. Finally, to help the plates survive wear and tear, they undergo a process called baking.

HYBRID plates are very durable with unmatched water ionization.

Myths About Platinum-Coated Titanium Plates

The biggest misconception surrounding platinum-coated titanium plates is the risk of titanium and platinum poisoning. However, it's just that – a misconception. Surgeons and dentists use implants made of these elements. Usually, high-quality alkaline ionized water machines use pure platinum titanium plates for the ultimate water ionization.

Furthermore, they do not disintegrate or corrode during electrolysis.

The cheap ones, especially from Taiwan and China, don't use them as they are expensive. Instead, they use Platinoridium, an alloy of platinum and iridium, which is 7 times cheaper than platinum titanium. Japanese made alkaline water filtration systems do not use iridium. It is not suitable for drinking water because it can have adverse effects on your health.

You also need to avoid any ionized water purifier that uses electrodes with white gold plating. It is not pure gold as it is an amalgam that consists of a minimum of 62% silver. In short, the best water ionizer for your whole house should have titanium plates coated with platinum.

Does the number of plates matter? Today, you will find natural alkaline water machines with 5 to 13 plates. The number of plates can determine the power consumption. When the electrodes use more power, the water ionization and the flow rate improves. You will have higher negative-ORP water for health benefits.

The more plates the water alkalizer has, the better for the water ionization. However, more plates means a higher price tag.

Another important factor is the size of theplate as many low budget ionzers claim to have many plates thus better electrolysation, but the fact is their plates are too small in size hence they would have less area of plates though high in number.

Inshort what matters in plate are:

1. Type of plate

2. Material and alloys used to make that plate.

3. Coating on the plate also whether its dipped or sprayed.

4. Number of plates

5. Size of plates

Types of Power Supply

There are 2 types of power supply used in an ionizer: switch Mode (SMPT) and Linear.

DC power supplies are available in either switch-mode (also called switching) or linear designs. While both types supply DC power, the method used to produce this power is different. Each type of power supply has advantages over the other one. Let's look at the differences between these two technologies as well as each design's respective advantages and disadvantages.

A switch-mode power supply converts the AC line power directly into a DC voltage without a transformer, and this raw DC voltage is then converted into a higher frequency AC signal, which is used in the regulator circuit to produce the desired voltage and current. This results in a much smaller, lighter transformer for raising or lowering the voltage than what would be necessary at an AC line frequency of 60 Hz. These smaller transformers are also considerably more efficient than 60 Hz transformers, so the power conversion ratio is higher.

A linear power supply design applies the AC line voltage to a power transformer to raise or lower the voltage before being applied to the regulator circuitry. Since the size of the transformer is indirectly proportional to the frequency of operation, this results in a larger, heavier power supply.

Each type of power supply operation has its own set of advantages and disadvantages. A switch-mode power supply is as much as 80% smaller and lighter than a corresponding linear power supply, but it generates high-frequency noise that can interfere with sensitive electronic equipment. Unlike linear power supplies, switch-mode power supplies are able to withstand small losses of AC power in the range of 10-20 ms without affecting the outputs.

A linear power supply requires larger semiconductor devices to regulate the output voltage and therefore generates more heat, resulting in lower energy efficiency. A linear power supply normally operates around 60% efficiency for 24V outputs, whereas a switch-mode power supply operates at 80% or more. Linear power supplies have transient response times up to 100 times faster than their switch-mode counterparts, which is important.

In general, a switch-mode power supply is best suited for portable equipment, since it is lighter and more compact. Because the electrical noise is lower and easier to contain, a linear power supply is roburst and better suited for powering ionizer analog circuity.

Which Electrolyser did I Choose?

So much depends on one's personal choices, lifestyle and requirements, I would always recommend doing your own research first. Personally, after my own research and experiments, I find Enagic's electrolysers to be ideal – not only because it's ahead of the competition in terms of quality and technology, but also because it's perhaps one of the most researched and widely studied products. Many esteemed medical practitioners recommend it too.

The water that comes out of enagics device Levelluk is called Kangen Water, which is the trade mark of Enagic company. Kangen water is Ionized Alkaline Water that significantly assists the body in creating a better balance. There are three different PH levels of alkaline drinking water, Clean Water, Beauty Water and Strong Acidic Water.

Alkaline Ionized Kangen Water is up to six times more hydrating than tap water. It's refreshing and provides superior hydration to the body.

Its highly alkaline qualities improve your pH balance, restore the body's balance and flush acidic waste. There are more powerful antioxidants in a glass of Kangen Water than a few pounds of blueberries. Every glass aids in weight loss as it flushes away harmful toxins.

Kangen Water is currently used worldwide by athletes, sports teams, Olympic athletes and bodybuilders. It is a rich source of extra oxygen, disposes off lactic acid, builds endurance and strength for workouts and competitions.

Kangen Water is used in homes, hospitals, disease control units, government agencies, medical and dental offices as well as restaurants worldwide. Forty plus years of technology have gone into creating what can be considered a highly superior water ionizer. Enagic is the only water purification and ionizer system that is medically accredited. They have four distinct certifications from the (WQA) Water Quality Association Worldwide, they are the only electrolyser in the world to have received Gold Seal Certification, it is the highest and most distinguished award for excellence and quality.

Some famous people who own Kangen water machines

International: The Arizona Diamond Backs, Jillian Michaels, Jennifer Lopez, Roger Dautry, Bill Gates, Brad Pitt, Angelina Jolie, Janet Jackson, Steven Seagal, UFC fighter Joe Stevenson, Jay-Z and Beyonce, Wade Lightheart, LA Lakers, Chuck Norris, Toby Keith, NY Yankees, US Olympics Ski Team, Demi Moore, Elton John, Santana, Magic Johnson, Jack Nicholson, and Chris Angel to name a few.

In India: Hrithik Roshan, Shilpa Shetty, Ajay Devgan, Sri Sri Ravi Shankar, Satpal Maharaji, Brahma Kumari (Mt. Abu), Vineet Jain, Sameer Gehlaut, Dr. Mahendra Prasad, Shahrukh Khan, Ekta Kapoor, Rajnikant, Iskon (Juhu), Akshay Kumar and many more.

Chapter 3

Why Use Ionizer

Water % of Human Body

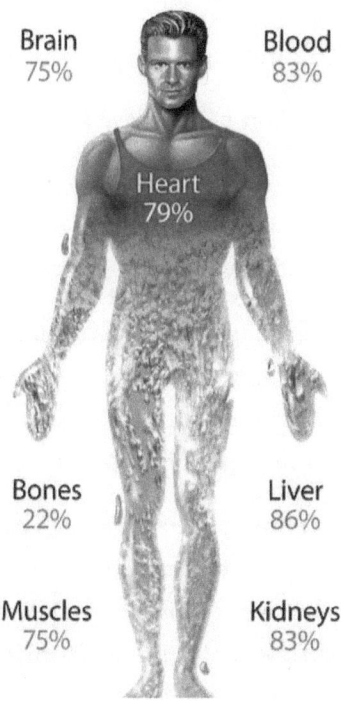

Brain
75%

Blood
83%

Heart
79%

Bones
22%

Liver
86%

Muscles
75%

Kidneys
83%

Our bodies are made up of 75% water. It's a well-known fact that staying well-hydrated is essential to our overall health and vitality. Water improves blood supply and circulation, ensuring that nutrients reach all the cells in all parts of the body. Moreover, it helps with digestion, detoxifies our system and keeps headaches and fatigue at bay. Drinking water regularly is also crucial to weight loss and maintenance.

But, here's the thing…

Not All Water Is Created Equal

The tap water that we rely on so blindly every single day of our lives, is not rich in the nutrients required for optimal health. In fact, it contains chemicals, pollutants and contaminants that continue to accumulate in the body and cause serious damage in the long run.

According to World Health Organisation:

- At least 1.8 billion people have access only to contaminated water.

- Contaminated water can transmit diseases such as diarrhoea, cholera, typhoid, dysentery and polio.

Just because the water flowing out of your tap looks clear, doesn't mean all's well with it. Especially, in a time when water scarcity continues to overshadow masses on a global scale. It is prudent to be proactive about something as basic and important as your drinking water!

Tap water quality varies from place to place and can contain toxins that even a electrolyzer filter isn't designed to eliminate. If a pre-filtration device isn't used, especially for hard water and water with high TDS (Total dissolved solvent), it could end up destroying the machine! The right pre – filter will not only add years to the life of your electrolyzer Water machine, but also to your life. A good pre-filter insures any quality problems that could arise due to various contaminants or other substances present in your water.

Most people get confused between the water purifier and the water electrolyser/ioniser, these are two different technologies with different uses, as the world suggest the job of a water purifier is to clean water and the ionizer makes your water healthly and as mentioned earlier ionizer should be installed after a pre-filter which is a water purifier.

How Does Ionized Improve the Quality of Water?

My considerable research has led me to understand that the real 'magic' of ionized water lies in the thorough ionization of tap water. You see, as water passes through an electrical field, it separates water molecules and creates what I consider to be the ideal antioxidant! The more water molecules are split, the more antioxidants are produced.

The greater the quantity of antioxidants contained in the water, the better the health results!

As we all studied in school, electrolysis is a technique where electricity is passed through electrodes composed of Titanium dipped in platinum, which have the ability to attract ions that conduct electricity. These ions that are naturally present in water are later concentrated into clusters of charged ions, either positive or negative, in turn leaving the original water cluster into a smaller charged hexagonal cluster. This process creates both Alkaline water (8.5 to 11.5 pH), and Acidic water (2.5 to 6 pH).

But a good electrolyzer creates five types of water, which are:

1. *Ionized alkaline Water* with pH 8.5 – 9.5 is perfect for drinking and healthy cooking. This hydrogen-rich water works to restore your body to an alkaline state, which is optimal for good health. It will enhance the flavours of vegetables and will cut down your usage of condiments and salts.

2. **Clean Water** with pH 7 is free of chlorine, rust and cloudiness. This neutral water should be used for preparing baby food, and taking medication so that your body easily absorbs the drug.

3. **Beauty Water** with pH 4.0 – 6.0 is not for drinking, as this water is acidic and is known for its astringent effect. It works wonders on the skin (toning and firming) and hair (detangling and conditioning). And also helps preserve the flavour of food like fish and shrimp if sprayed on before freezing.

4. **Strong Acidic Water** with pH 2.7 or less contains disinfectant properties. Use this water to sanitize a place or object, and prevent cross-contamination.

5. **Strong Alkaline Water** with pH 11 or more is ideal for cleaning. This water has dissolving and heat conducting benefits. It can be used to thoroughly clean cutting boards, dishcloths, oil deposits as well

   as for general cleaning in the kitchen. As it has extra absorption power it will remove coffee, soy sauce and oil stains with ease, while you use less detergent for your dishes and consume a third of the water.

How & Why It Works

Biological medicine focuses on enhancing and replicating the normal functions of the body. When we support and encourage the natural rhythms and cycles of a healthy body we begin to reprogram the body's systems to function properly again. My goal was to find a way to naturally enhance the body's entire digestive process from the stomach through the large intestine.

So I followed my old protocol but substituted various pH levels of ionized water and the results were remarkable!

I used a 2-step approach to increasing the normal production of hydrochloric acid in the stomach.

- **First**, eliminate or neutralize it so that the system is forced to start production of the acid just in time for food to arrive! I used small quantities of pH 11.5 ionized water to accomplish this task.

- **Second**, add a bit more acid to the stomach to ensure the early stages of digestion are complete. To accomplish that I used just a little pH 2.5 water, because it mimics a diluted, non-toxic hydrochloric acid.

Most people don't know that conditions like acid reflux, GERD, and the old fashioned heartburn occur when there is not enough acid in the stomach. You see, there is a sphincter between the stomach and the esophagus that is designed to close when adequate amount of stomach acid is released.

So, it is essential to be certain that your stomach has all the acid it needs to break down the food you are eating and signal the sphincter to close.

I also used a 2-step process to adjust the pH of the small intestine, to encourage the proper flow of bile and the release of pancreatic enzymes to complete the break down of fats, proteins and carbohydrates during the final stages of digestion.

- **First**, by eliminating the acids in the stomach, alkaline drinking water is allowed to pass through the stomach directly into the small intestine without interference. Since water is the only substance we ingest that is not digested it will pass virtually undisturbed through the stomach and into the small intestine where it is able to create an alkaline environment.

- *Second*, by consuming a large amount of alkaline antioxidant-

- rich drinking water at one time, there is an adequate amount of both water and antioxidants delivered that can be absorbed

- immediately into the blood stream and the lymphatic system. This encourages the body to eliminate toxins that are responsible for creating unhealthy thickening of the bile, preventing it from freely flowing into the small intestine.

Utilizing this Water Protocol consistently creates an ideal pH for complete digestion, encourages the normal production and timely release of stomach acids, bile as well as pancreatic enzymes, while retraining the system to work properly on its own. Complete digestion is essential for overall health and natural self-healing.

How Ionized Water Can Change Your Life…

There are numerous scientific studies that have been performed across universities and hospitals worldwide to prove the four overarching benefits of drinking restructured ionized water on a regular basis:

1. Increase in hydration

Ionization reduces the size of water molecules by two-thirds of its original size. This reformed shape and size of water molecules makes absorption easier, enhances tissue repair and improves waste removal.

2. Balances the body's pH level

Ionized water has a higher pH level due to the splitting up of water molecules which give rise to OH-ions and ionic alkaline minerals.

These ions increase the bicarbonate buffers in the blood, thereby balancing and neutralizing the body and excreting acids and toxins.

3. Improves Oxygen supply to the blood

Fresh restructured ionized water is rich in hydroxy ions that donate free electrons to unstable oxygen free radicals. This, in turn, stabilizes the oxygen molecules which are better than the free radicals, as they are non – reactive, improve mental alertness and energize the body.

4. Neutralizes free radicals

Research tells us that active oxygen molecules are free radicals that can damage normal tissue and speed up aging and deterioration of internal systems. Free radicals are also what speed up the aging process. In order to reverse this damage, it's important to neutralize free radicals.

Restructured ionized water turns these active molecules into inactive ones, making it a potent antioxidant.

Ionized Water Aids In

- DISCHARGE TOXINS & EXCESS BODY FAT
- STABILIZE BLOOD SUGAR & INSULIN
- NORMALISE BLOOD PRESSURE
- REMOVAL OF ABNORMAL GASTROINTESTINAL PUTREFACTION
- MAINTAIN A HEALTHY COLON
- RESOLVE URINARY TRACT INFECTIONS
- REDUCE CANDIDA & FUNGUS PROLIFERATION

- MINIMISE CHRONIC PAIN

- ENHANCE THE BODY'S HEALING POWER

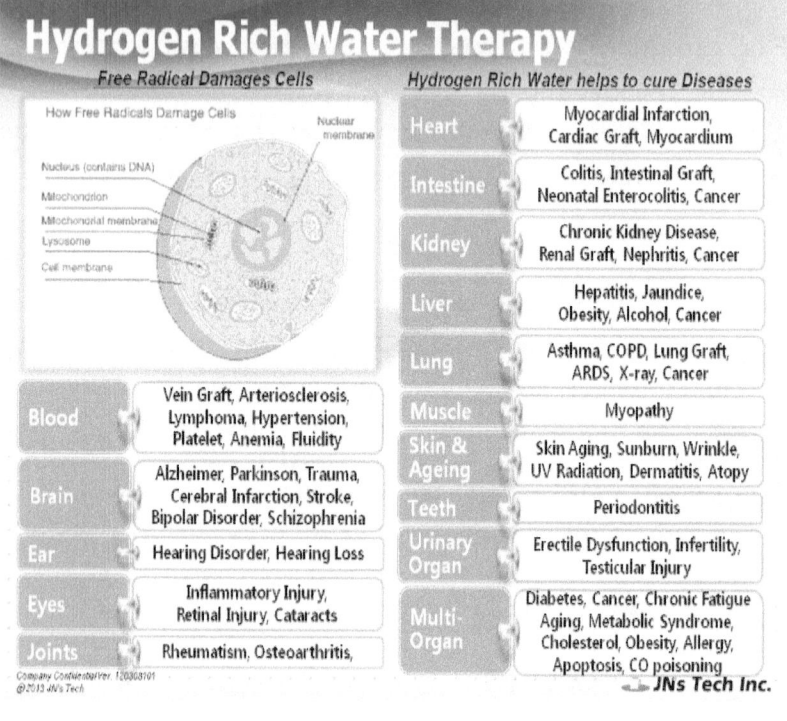

Chapter 4

Why be Alkaline

Alkaline

While a myriad variety of diets exist, none are as conducive to a long healthy life free of disease as an alkaline diet is.

pH of the gastrointestinal tract

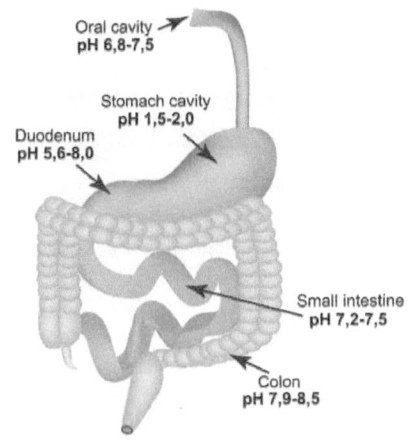

Oral cavity
pH 6,8-7,5

Stomach cavity
pH 1,5-2,0

Duodenum
pH 5,6-8,0

Small intestine
pH 7,2-7,5

Colon
pH 7,9-8,5

In 2012, the Journal of Environmental Health published a review that found that balancing the body's pH through an alkaline diet can actually be helpful to reduce morbidity and mortality caused by numerous chronic diseases, ailments and deficiencies, such as hypertension, diabetes, arthritis, deficiency of Vitamin D and low bone density among others.

An alkaline diet consists of food and water that's highly alkaline in nature, such as fresh vegetable, fruits and unprocessed plant-based proteins. This kind of diet protects healthy cells and balances essential minerals. It also prevents plaque formation in blood vessels, stops accumulation of calcium in urine, prevents kidney stones, reduces muscle wastage and spasms, build stronger bones and so on.

Over the last 200 years, the food we've been consuming has seen a steady decline in potassium, magnesium and chloride levels and a steady increase in sodium levels. While normally it is the kidneys' job to maintain electrolyte levels, when we're taking in highly acidic substances, these electrolytes are expended to control this acidity. Over the course of years our diet has evolved to include refined fats, simple sugars, sodium and chloride, which in turn have increased metabolic acidosis. This accelerates the aging process, gradual loss of organ functions, and degeneration of tissues and bone mass.

Best Alkaline Foods:

- Fresh fruits and vegetables such as mushrooms, citrus, dates and raisins, spinach, tomatoes, avocado, black radish, alfalfa grass, barley grass, cucumber, kale, jicama, wheatgrass, broccoli,

- oregano, garlic, ginger, green beans, endive, cabbage, celery, red beet, watermelon, figs and ripe bananas. Include as many raw fruits and vegetables in your diet as possible.

- Plant proteins such as almonds, navy beans, lima beans, and most other beans and legumes.

- Alkaline water with a pH of 9 to 11. Adding lemon or lime to this water and drinking throughout the day can also boost alkalinity.

Foods to Avoid to Maintain Alkalinity

The following foods are notorious for their high acidic properties and anti – alkaline tendencies.

- High-sodium foods: Processed foods contain tons of sodium chloride. This includes table salt, excess consumption of which constricts blood vessels and creates acidity.

- Cold cuts and conventional meats

- Processed cereals (such as corn flakes)

- Eggs

- Caffeinated drinks and alcohol

- Oats and whole wheat products: All grains, whole or not, create acidity in the body.

- Milk: Calcium-rich dairy products are some of the highest contributors to osteoporosis. That's because they create acidity in the body! When your bloodstream becomes too acidic, it will steal calcium (a more alkaline substance) from the bones to try to balance out the pH level.

- Peanuts

- Pasta, rice, bread and packaged grain products

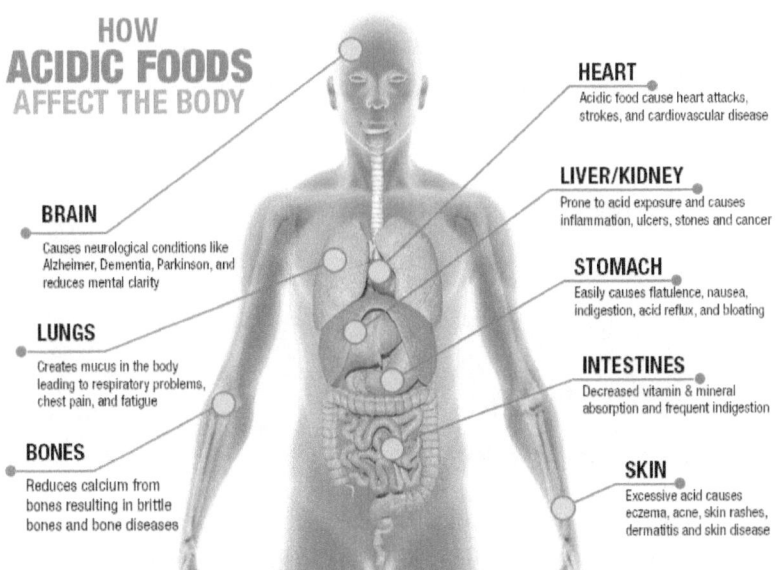

HOW
ACIDIC FOODS
AFFECT THE BODY

HEART
Acidic food cause heart attacks, strokes, and cardiovascular disease

LIVER/KIDNEY
Prone to acid exposure and causes inflammation, ulcers, stones and cancer

BRAIN
Causes neurological conditions like Alzheimer, Dementia, Parkinson, and reduces mental clarity

STOMACH
Easily causes flatulence, nausea, indigestion, acid reflux, and bloating

LUNGS
Creates mucus in the body leading to respiratory problems, chest pain, and fatigue

INTESTINES
Decreased vitamin & mineral absorption and frequent indigestion

BONES
Reduces calcium from bones resulting in brittle bones and bone diseases

SKIN
Excessive acid causes eczema, acne, skin rashes, dermatitis and skin disease

Habits to Avoid to Maintain Alkalinity

Regular, seemingly harmless habits that you do on a day to day basis could be causing a severe imbalance in your pH levels. These include.

- Overconsumption of alcohol, drugs and caffeine

- Overconsumption of antibiotics

- Overconsumption of artificial sweeteners

- Chronic stress and fatigue

- Declining nutrient levels in foods due to industrial farming

- A diet that's low in fibre

- Too much as well as too little to no exercise

- Overconsumption of non-grass-fed animal meat

- Excess hormones from foods, health and beauty products, and plastics

- Exposure to chemicals and radiation from household cleansers, building materials, computers, cell phones and microwaves

- Food colour and preservatives

- Pesticides and herbicides

- Pollution

- Poor chewing and eating habits

- Shallow breathing

Why Ionized Water Works?

70% of our body is made up of water! Also, water composes a massive of 75% of all muscle tissue, 80% of our blood and 90% of our brain.

Not just that, even all the biochemical reactions that occur in the body depend on water. Here's a fun fact, normal metabolic activity can occur only when cells are at least 65% water bodied.

The way Ionized Water works is that it scavenges for free radicals and provides a way for toxins to exit the body by neutralizing the ions. A fair percentage of toxins are stored in fat cells, tissues, organs and lymphatic fluids. Ionized water flushes these out, thus helping in fat loss as well.

Ionized Water is rich in ionic minerals such as calcium and iron, and its molecules get absorbed six times faster than tap water or bottled water, thus ensuring faster absorption of essential nutrients.

Due to its alkaline nature, Ionized Water can balance a system that's constantly suffering from acidity, which in turn causes pain and inflammation. To offset this, the body dips into the alkaline minerals stored in the bones and tissues, which further causes erosion of bone density.

This is where Ionized Water comes to the rescue and helps the body preserve and maintain its mineral reserves.

Better Than Bottled Water...

As much as bottled water may brand itself as reinforced with essentials vitamins and minerals, it still falls short of enriching the body. Moreover, specialty water isn't exactly financially viable – most bottled waters typically sell for about Rs.20 or more per litre. Compare that to your body's daily needs of about 3 to 4 litres per day, if you were to go for bottled water, it would probably cost you Rs.60 to Rs.80 a day! If you're to follow this pattern for 15 years, you would end up spending close to Rs.4,38,000 on water alone. But, have you experienced any of the promised benefits?

On the contrary, with Ionized water, you have your very own fountainof youth at your disposal, that makes nutritionally enriched accessible to yourself as well as your family.

Moreover, using Ionized Water minimises damage on the environment by reducing your use of one-time-use plastic bottles.

What about using tap water as an alternative?

Tap water is known to contain up to 90 legally permissible chemicals! Chlorine and various by-products of chlorine have been found in tap water. Chlorine in water has been linked to cancer in the bladder, breast and other parts of the body. Also, the chlorine that is used to kill bacteria in tap water is known to kill healthy bacteria in your gut as well.

It is best that one invests in a water filter than consuming water directly from the tap. But, purification is just a tiny part. Restructured ionized water has a set of benefits that are incomparable to a simple water filter!

Experts Speak

"I have administered over 5,000 gallons of this water for about every health situation imaginable. I feel that restructured alkaline water can benefit everyone."

– Dr. Theodore Baroody
Author, Alkalize or Die

"Alkaline water rids the body of acid waste. After carefully evaluating the results of my advice to hundreds of individuals, I'm convinced that toxicity in the form of acidic waste is the primary cause of degenerative disease."

– Sherry Rogers
MD Author, Detoxify or Die

"Drinking alkaline water is a great way to neutralize and flush out all of the toxins and acids that drain from the tissues and bodily fluids, and to quickly rehydrate the body and keep the blood alkaline."

– Daniel Reid,
Author, The Tao of Detox

"By drinking alkaline water, the aging process can be reversed and the wastes can be reduced in the long-

term to a level of a much younger person. The functions of the organs can be revived."

– Harald Tietze
Author, Youthing

The Importance of Water In Numbers

- Some experts estimate that up to 75% of Americans are dehydrated to the extent that it affects their health.

- The main cause of daytime fatigue is simply lack of water.

- Minimum water intake is 1/2 ounce per pound of body weight.

- As little as a 2% drop in the amount of water in your brain is responsible for mental confusion such as short-term memory loss, being unable to focus and forgetting how to do simple math calculations.

- Many people think they are hungry when they are actually thirsty. A University of Washington study showed that hunger pangs can be relieved by drinking one glass of water in 98% of the dieters surveyed.

- Research shows that about 8-10 glasses of water a day may significantly ease back pain and joint pain for up to 80% of sufferers.

- Drinking at least 5 – 16oz. glasses of pure water a day decreases the risk of getting breast cancer by 79%, colon cancer by 45% and bladder cancer by 50%.

Clinical research has documented the effect of restructured ionized water on subjects with chronic constipation. When constipation is prolonged, it impacts other body processes and health. According to Dr. Mu Shik Jhon, consumption of hexagonal water improved transit time and frequency of bowel movements. Other research has documented the effect of Ionized Water on the prevalence of abnormal gastro-intestinal putrefaction.

Signs You Need Ionized Water

When your cells, tissues, and interstitial fluids are overly acidic, you are more likely to experience the following effects:

- Tire easily and become fatigued

- Find it more difficult to think clearly (brain fog)

- Develop a pessimistic outlook on life

- Lack the energy and vital spark to achieve your goals and aspirations

- More frequent colds, flu, allergies and respiratory ailments

- Stiffness, joint pain and arthritis

- Chronic fatigue, chemical sensitivities or fibromyalgia

- Chronic long-term medical problems like high blood pressure, autoimmunedisorders, cancer, heart disease, diabetes, or inflammation

- Retain toxins and heavy metals

When a body is acidic, it holds onto heavy metals such as mercury, lead and cadmium. These heavy metals, in turn, create high oxidative stress that further acidifies the body. Heavy metals are the cause of many degenerative conditions.

Restructured ionized water has been shown to have an immediate detoxifying effect on the body because of its tendency to bind with acids and toxins, including heavy metals. When first drinking restructured ionized water, toxic individuals may experience classic detoxification symptoms such as itching, body aches, and headaches.

How Much Water do I have to DRINK

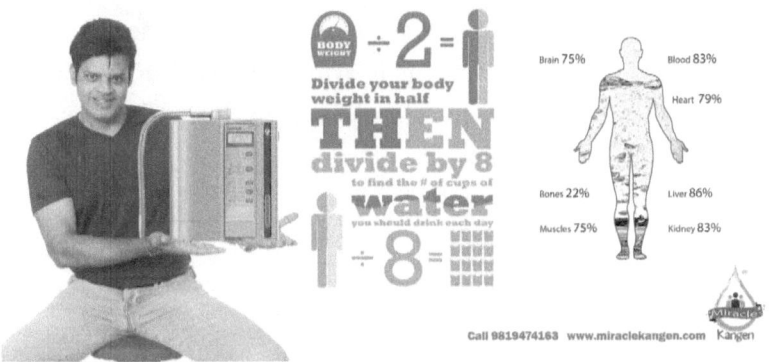

It's common to think that dehydration is something that would happen if we got lost in a desert or had to go without water for several days. The truth is – chronic dehydration is widespread and more common than we think. At a certain level, most individuals are dehydrated.

Unlike camels, the human body cannot store water to sustain for days and weeks. Dehydration occurs when the body's stock of water isn't replenished. Chronic dehydration occurs when small amounts of water is not replaced day after day. Drinking caffeine or alcohol only adds to the problem. These substances cause a net loss, rather than a gain of water in your body. Those who regularly drink caffeinated or alcoholic beverages are literally "drinking themselves dry."

According to Mayo Clinic, the average adult loses more than 10 cups of water every day, simply by breathing, sweating and passing urine. When the body is depleted of water, numerous functions begin to slow down. You end up experiencing headaches, constipation, brain fog, afternoon fatigue and many other ailments. Persistent dehydration leads to water rationing, and long-term water rationing can cause premature aging and a variety of diseases. Water rationing allows the body to retain water for essential, life-sustaining functions that cannot be put 'on hold'. Cells need some amount of water to function optimally. Cellular water is reduced during rationing, and each cell therefore functions below optimal levels. This results in symptoms that are mistaken for illness. Asthma, arthritis, blood disorders, digestive disorders (including heartburn), and many chronic pains are often an indication of chronic dehydration and water rationing.

Chapter 5

Wellness & Ionized Water

Ionized Water and Cancer Treatment

According to a study conducted by Webster Kehr at the Independent Cancer Research Foundation, alkaline water has a high redox potential that help inhibit the spread of cancer and aid in killing cancerous cells. Although the treatment cannot be recommended as a standalone plan in fighting cancer, it works as a potent supplementary method.

Ionized water has antioxidant properties that neutralise free radicals, making more oxygen molecules available to cancer cells. These molecules either slow down the growth of cancer or kill the cells entirely. Ionized water creates an alkaline environment in the body, which is not conducive for cancer cells. When the blood becomes increasingly acidic in nature, excess acidic molecules get deposited somewhere in the body. Over the course of time, the areas where these acidic molecules are stored deteriorate and the cells begin to die. Surrounding cells survive this by becoming abnormal, also known as malignant cells. These malignant cells fail to respond to the brain's commands. As a result, they begin to grow out of control, causing cancer. Modern medicine treats these cells as bacteria or viruses using chemotherapy, radiation or surgery. Unfortunately, these methods do nothing to combat the acidic environment within the body. On the contrary, across Asia, alkaline or Ionized water is served to cancer patients as part of regular treatment, to bring the body's acidic environment under control.

In addition to drinking alkaline water, you may also bathe in it. It will stimulate oxygen supply throughout the body without having to go through the digestive tract. All you have to do is add a gallon or more to your regular bath water.

The Toxins in Our Water, Cancer, Disease & Ionized Alkaline Water Plus

If you're already afflicted by cancer or another disease, it's important to know and remember that even if you cure it, the disease it likely to return unless you change what made you vulnerable to the disease in the first place. Another important fact to remember is that Alkaline Ionized Water is NOT a cure, it's merely a facilitator of change – it rewires your cells to function smoothly, optimally and achieve peak health. But, most diseases are preventable!

By drinking alkaline ionized water, you can help counteract the acids in the foods and drinks you consume. By having more balancing pH buffers [and antioxidants] in your body, you can then increase your chances for maintaining optimal health. Every day, hundreds of thousands of people are drinking alkaline ionized water and are changing their futures! They're theoretically slowing down the aging process, promoting healthy weight loss, detoxifying their bodies, boosting immunity against disease and increasing the body's ability to absorb nutrients, vitamins and minerals.

Who knows how many diseases are being prevented? I actually will never know the answer to that question, but I do know that the amount of energy and healthiness that I feel every day [without fail] is telling me something! Since I've been drinking ionized water, I am factually healthier. I am healthier now, at 60-plus, than I was at 40!

Another major factor that acts as the root of most health problems and is often ignored is dehydration. Common effects of dehydration can include: low energy, migraines, Type II diabetes, hypertension, weight gain, low back pain and colon problems. I'll bet the last thing 99% of the population thinks when they get a headache, back pain or heartburn is that they should go drink water. We are often trained to not like water, take it for granted or completely disregard it. Positively charged tap water is not very hydrating, and I believe people intuitively know that. Ionized water is clearly different.

This powerful alkalizing, antioxidant water provides superior hydration to your cells. Alkaline ionized water filters [water ionizers] reduce the size of the water molecule clusters, resulting in significantly better hydration per glass, which means a lot more oxygen can reach and actually be used by your cells. The antioxidants in ionized water are the most powerful there are.

Ionized Water and Diabetes

Ionized water is medically known to help the human body recuperate from dehydration as well as other types of diseases that are associated with dehydration.

Diabetes Type 1 and Type 2:

What you need to know?

Diabetes can be classified into two types:

Type I which are known as "Insulin – dependent" diabetes its strikes children and very young people. It most often occurs in young people. They have pancreases that make little to no insulin. They are almost always insulin dependent for life.

The second type is Type 2 Diabetes; this is adult onset diabetes. Type 2 can be hereditary or can occur because of poor diet and fluctuating sugar levels, which exhaust the pancreas.

This is the most widely spread form of diabetes in the United States. Though more easily treated and manageable than Type 1.

Ionized water's qualities of smaller water molecules hydrate cells faster and flushes toxins and lactic acid. It also carries ionized calcium to the pancreas.

Doctors in Korea and Japan have been able to reduce or eliminate the need for insulin with their patients just by using Ionized Water. By combining the Ionized water with a healthy Alkaline based diet and foods they have successfully treated patients with Type 2.

The reason the body is better managed by using Ionized water comes down to its PH levels. The pancreas secretes insulin which is high in alkaline. Insulin helps lower blood PH. Since Ionized water has smaller molecules. Its ability to flush toxins is greater and hydration are better. The ionized calcium is easy for the pancreas to use and is necessary for it to do its job better.

What does this mean for diabetes? Alkaline ionized water, along with a healthy diet and regular regimented exercise has proven in a variety of studies to reverse Type 2 diabetes without the use of medications. One study, done in a hospital in Korea, divided diabetics into two separate groups. The first group continued their normal routine along with their Insulin injections. The second group, under close supervision, stopped taking their Insulin injections and started drinking only ionized alkaline water.

In his book, "Water of Life", Dr. Won H. Kim, professor at the Medical School at Yonsei University, reports of this study, "in less than one month the blood sugar level of the group drinking alkaline water had amazingly decreased in contrast to that of the insulin-injected group. Daily fluctuations in blood sugar continued to occur in the case of the insulin group, while those drinking the alkaline water maintained remarkably stable levels."

By removing the build-up of toxins in the body, Ionized Water prevents the ketones from building up. The ketones are then excreted through the urine most especially when it is impossible for the blood to handle these ketones. This condition can be very dangerous to people diagnosed with diabetes. Continuous occurrence of this condition may eventually lead to coma.

Unstable pH levels in the body can be very unhealthy as it may cause the system and organs to malfunction. Certain studies also proved that Ionized Water can actually aid in keeping the pH levels stable. Moreover, Ionized

Water also aids in melting the acids from the foods that people eat. People with a diet that's prone to make them obese are recommended Ionized Water to improve their health.

A very good antioxidant, Ionized Water also functions as a nutrient or vitamin that provides an extra electron that functions to add necessary radicals in the body. When the electrons in the radicals are back, health and vitality are restored again.

By combining Ionized Water with a healthy diet plan, proper exercise and effective medical care, a person can reduce their chances of developing this disease. However, if a person is already diagnosed with Type 2 diabetes, the simple act of adding Ionized water to their current health regimen can reverse the effects and eliminate or reduce future health risks associated with the disease.

Here's How Water Ionizers Help The Stomach and Colon

Alkaline Ionized Water benefits the stomach too!

Some people get confused about the stomach and think it always wants to be acidic. That is far from the truth. The stomach does need to be acidic for meals and if you add a squeeze of lemon to your alkaline ionized water at meal-times it will not upset that balance. I don't usually drink with my meals, so that's my solution for meals.

In-between mealtimes, the stomach likes to be alkaline. The water we drink in-between meals (as long as it's slightly alkaline) has the capability of absorbing into the rest of our body right through the walls of the stomach.

As a matter of fact alkaline ionized water improves digestion. Throughout the entire length of your intestines, where important phases of digestion occur, it is healthier to have slightly alkaline water. Good bacteria in the gut prefer slightly alkaline water. So, during the between-meals times, one should just sip on alkaline ionized water all throughout the day.

Among the benefits of ionized water are improved conditions in the GI tract including better elimination, digestion, energy, and overall health. Water has no calories, and so it doesn't trigger digestion or reduce the stomach acids at all. It just absorbs into your body or goes straight through to the intestines in a couple of minutes.

The stomach can adapt to all pH-levels of alkaline ionized water. It is weakly alkaline (meaning it can neutralize with only a drop of HCL) and does not pose a threat to digestion at all.

Stomach

The stomach is the only organ in the entire digestive system that requires a low acid pH level in order to do its job, and it produces it all by itself by injecting HCL or hydrochloric acid, into the stomach as food arrives.

The stomach, when dormant, will only have a tiny bit of acid in it, but when digestion is stimulated gastric acid, containing mostly HCL, is produced on demand to handle the load. HCL's pH is 1-1.5.

Organisms in the food are likely to be killed by the low acid of the stomach juices. This makes it part of the body's immune system.

The acid itself doesn't break down the food, but rather the low pH environment is conductive to the enzymes doing their job to break down the protein walls.

Does Alkaline Ionized Water Upset HCL Balance in The Stomach?

No. HCL is a strong acid, creating a low acid pH in the stomach during digestion, but unless you're actively eating, there isn't much of it [or much need for it] in the stomach. Water, by itself, does not trigger the

production of HCL or digestion, so is allowed to pass through the stomach quickly and undigested without upsetting the overall pH balance of the stomach.

Small Intestine

Maintaining an alkaline balance in the intestines can make a great contribution to a person's colon health as well as their overall well-being.

The small intestine plays a significant role in the absorption of nutrients from the food. The mucosa of the small intestine is about 8.5 pH. All the way along the intestines, more enzymes are added and more food is absorbed until, by the time the chyme i.e. digested food, reaches the large intestine 95% of the nutrients should have been absorbed. The nutrients go into the blood, then straight to the liver for filtering before entering the regular circulatory system.

Ionized water is very beneficial in the small intestine for many reasons.

It assists the small intestines to maintain a healthy 8.5 pH so as to inhibit the growth of parasites, fungii, and other undesirable microorganisms. Parasites going unchecked are often a large contributor to malnourishment and weight problems.

Proper pH stimulates the enzymes to digest whatever food hasn't been broken down thoroughly enough by now.

Ionized water, being absorbed at this point in the GI tract, is healthy for you in many ways. The antioxidants will go to the liver and help to detoxify it [then will go to the rest of the cells in the body to do the same]. Also, the micro-clustered water will enable better absorption of water into the body and so keep the body well-hydrated. This will put less strain on the large intestine for water, and so will help stools be softer, and move faster through the colon.

The Large Intestine

The large intestine's main job is to absorb water into the body, produce Vitamins of B and K, absorption of a few more nutrients, and lastly – quick elimination of waste.

A good, healthy, alkaline pH level in the large intestine helps to inhibit the growth of bad bacteria and maintain the proper balance of good bacteria [intestinal flora].

Note: Insufficient water or oil in the diet or insufficient intestinal flora "good bacteria" can cause the stool to be hard, sticky and trapped in the walls of the large intestine.

If our bodies don't eliminate waste properly, toxins get reabsorbed from our bowels. Alkaline ionized water helps to promote good elimination through the colon.

Dr. Shinya is the inventor of the colonoscopy and has learned many good tips for keeping a healthy GI tract. In addition to drinking plenty of good water, Dr. Shinya advises we develop eating habits that are good for our colon and GI tract health. He also advocates daily exercise, rest/sleep, and managing our stress better.

The following is the "Good Eating Habits" advice from Dr. Shinya's website:

Good Eating Habits: 90% fruits and vegetables and 10% protein. Dr. Shinya's points for good eating are:

Eat unpurified grains or cereals.

Eat more vegetables. More seafood, less meat.

Eat raw foods.

Do not eat oxidized foods. [i.e., over-cooked, hydrogenated, or spoiled] Eat fermented foods.

Avoid milk and milk products. Take Vitamins and Minerals

Be disciplined with the food you eat.

How else can you take causative action to improve digestion?

Drink ionized water: Drinking alkaline ionized water will help establish and maintain a healthy, necessarily slightly-alkaline, balance of pH in the GI tract as described above.

Combine foods properly. Try to not mix sugary foods, refined carbohydrates or refined starches with meals.

Avoid junk food. This just gives you empty calories, taxes your digestive system, toxifies the liver and other organs of digestion, slows down digestion, and can contribute to malnourishment and obesity.

Eat unsaturated fats [cold-pressed oils for salads and olive oil for cooking] rather than trans fats and saturated fats. Good fats help to supply you with quick energy and make you feel satisfied, so as to have fewer cravings in general. Your body needs fatty acids, but only the good ones. Bad fats contribute to hunger, overeating, malnutrition, heart disease, and many other ailments.

Don't overeat. This triggers insulin, which interferes with your body's sugar/energy balance and makes you feel more tired [and hungry] later.

Stress can lead to poor digestion and an overly acidic body. Therefore, it is suggested that you adopt a healthy, mindful lifestyle. Stress-free doesn't imply "action-free" or a lethargic existence. Quite the contrary, the more productive and purposeful a person is, often the less stress they usually experience.

Move your body frequently and periodically exert yourself, so as to keep your muscles toned, use up your calories, improve circulation and improve digestion.

Lose weight if you need to. Overweight people demand more calories, get hungrier, and so tend to eat more. The more you eat, the more difficult it is for your digestive system to function properly.

Ionized Water and Sports Performance

Competitive, elite athletes and sports trainers know that subtle changes in pH can have profound effects on overall health, feeling of wellness, level of fatigue, pain, weight, ability to train and athletic performance. When we exercise, the increased use of muscle glycogen for energy produces lactic acid, pyruvic acid, and CO_2, which decreases muscle pH. The harder you exercise, the quicker your muscles become acidic which leads to fatigue. Accumulation of acid also limits production of ATP, the energy molecule, and disrupts enzyme activity that produces energy. When muscle pH falls below 6.5, it stops working altogether. Acidity also reduces muscle power directly by inhibiting the contractile action of muscle fibres.

As the body metabolizes food, acid waste is created which must be removed or neutralized through the lungs, kidneys (urine) and skin. Athletes, coaches and practitioners of holistic and traditional medicine are paying more attention to this area.

The use of restructured ionized water is proving to increase competitiveness and overall performance in world class athletes.

A diet that supports alkalinity is also recommended by sports nutritionists. Consuming restructured ionized water will reduce the accumulation of acidity in exercising muscles, improving workout intensity and recovery time.

The benefits of the alkaline water created through electrolysis far exceed just its ability to gently raise the pH of the cells and tissues of the body and to neutralize acids. Because alkaline water has gained a significant number of free electrons through electrolysis, it is able to donate these electrons to active free radicals in the body, thereby becoming a super antioxidant. By donating its excess free electrons, alkaline water is able to block the oxidation of normal tissue by free oxygen radicals.

Another significant benefit of electrolysis is that the cluster size of the alkaline water is reduced by about 50 percent from the cluster size of tap water. This allows ionized alkaline water to be much more readily absorbed by the body, thereby increasing the water's hydrating ability, its ability to carry negative ions and its alkalising effect to all the cells and tissues of the body.

Most people, including most athletes, do not consume enough alkaline mineral-rich foods – such as nuts, fruits, and vegetables. Instead their diet contains high amounts of acid-fanning foods – such as meat, fish, poultry, eggs and dairy. Because of this dietary imbalance, they may be at risk for increased acidosis that affects overall health and sports performance. Since proper hydration is also a key factor in preventing exercise fatigue, consuming restructured ionized water before, during and after exercise can help.

Ionized Water and Your Lifestyle

Our diet has changed drastically in a short time period. The average person's diet contains a preponderance of acid-forming foods such as meat, poultry, dairy products, some fruits, nuts, refined sugar, com sweeteners, artificial sweeteners, chocolate, refined flour products, soft drinks, beer, wine, coffee and black tea!

Regular and Diet sodas are probably the most acidic food people consume, at a pH of 2.5. Most people do not have enough alkaline buffer reserves to offset or neutralize the acid waste produced by consuming the Standard American Diet – also known as SAD – composed primarily of these foods. Less acid-forming foods include vegetables, starches, non-gluten grains, legumes, seafood, eggs and certain fruits.

Experts recommend that 80% of the diet should be fresh, organic fruits and vegetables, raw if possible. Soaked almonds are also a good alkalizing food, and asparagus is one of the most alkaline – forming vegetables. If you eat meat, make sure it fits into the 20% of your food intake that is reserved for acid-forming foods.

Limit servings to no more than 3 oz. portions of organic, range fed poultry, red meat, or clean, mercury-free fish. Avoid dairy and fatty meats, limit your intake of gluten grains and avoid acidic drinks like soda, sports drinks, coffee, tea. beer, wine or alcohol. And drink plenty of restructured ionized water!

Acid Accumulation and Stress

Acid comes from three sources – food, pollution, and stress. Of these three, stress is the greatest problem. One surge of adrenaline produced in response to a fight or flight stimulus can negate the benefits of an alkaline diet. So, stress management as well as diet management is essential to maintaining an alkaline body. Think about it. We work 40 to 50 hours a week dealing with constant "fight or flight" stimuli with hardly any breaks to calm ourselves down. We consume fast food and coffee – both acid – forming – for quick bursts of artificial energy to just get through our work day. Then we come home to family stress, household chores, and financial obligations we can't quite meet. We never really relax and give our bodies a chance to neutralize all the acid we produced through stress and from eating very acidic foods. Instead, we consume a beer, glass of wine or alcoholic drink, a soda, sweetened juices or sports drinks, and acid-forming snacks, adding to our acidic overload.

Acid builds up in our bodies over time. We might not notice much in our 20s and 30s, but in our 40s and 50s we begin to show symptoms of acid imbalance – digestive problems, headaches, obesity, bone pains, elimination issues, muscle tension and pain, heart problems, high blood pressure, diabetes, arthritis and more.

Cooked foods have a deficit of electrons, a positive ORP and an abundance of positive ions. They are dry and dehydrating. They also acidify and add toxins to the body. All these qualities lead to disease and therefore are the exact opposite of what we should put in the body. Ionized water mimics many of the same attributes in nature that bring us health. Ionized water has the same characteristics as raw foods:

- An abundance of free electrons

- Alkaline pH

- A negative ORP – antioxidant properties

- Hydrating and detoxifying effect

Chapter 6

Understanding Various Water Uses

Specific pH levels of Ionized Water can be used to tackle various problems pertaining to your physical, physiological and emotional well-being.

11.5 Water – Strong Alkaline

Strong Alkaline Water has an Oxidation Reduction Potential (ORP) of – 700 mV to – 850 mV. It's extremely effective in combating free radicals and can greatly reduce inflammation. Those suffering from arthritis or related inflammatory conditions should do a Strong Alkaline Water compress.

On another small, but significant note, Strong Alkaline Water can also help combat acne, eczema and psoriasis. Where outbreaks are sore and irritated, I recommend a quick spray of Strong Alkaline Water to neutralise any bacteria. Leave it on for 5-15 minutes; follow with Beauty Water. I have seen large pimples literally disappear within hours by dabbing them with Strong Alkaline Water. Wrap sunburned areas in a cloth soaked in Strong Alkaline Water and leave for 15 – 20 minutes. Follow with Beauty water and let dry.

Strong Alkaline Water is also best kept in a dark bottle in the refrigerator.

- *Good night's sleep:* Drink 1/2-1 ounce of 11.5 before bedtime to help release Melatonin for sound sleep

- *Eye wash:* Use an eye cup to rinse eyes. Upon removing the eye cup from its packaging, soak in Strong Acid Water (pH 2.5) for 1-2 minutes to clean and disinfect. Rinse the eye cup thoroughly with Strong Alkaline Water (pH 11.5) Fill the cup, following package directions, with Strong Alkaline Water (pH 11.5) Place cup firmly around one eye, keeping your eye open, tilt your head back and gently roll your eye as though you were attempting to look up, down, and from side to side. Continue this for approximately 1 minute. Now that you have completed one eye, toss the water away. Rinse the eye cup thoroughly with Strong Acid Water (pH 2.5) Repeat steps for your other eye. To maintain healthy eyes, follow this protocol 1-3 times per week. When working to improve any eye condition, follow this protocol at least 2 times daily and up to 10 times per day.

- *Grease in eyes:* Spray 11.5 as needed to soothe and heal eye.

- *Eye makeup remover:* Spray on eyes to dissolve and remove make up.

- *Puffy eyes:* spray on eyes to reduce puffiness, A small misting bottle works well or soak gauze pads in 11.5 PH place on eyes for 10-15 minutes, let air dry and spray face with beauty water.

- *Hot bath soak:* Use one gallon of 11.5 added right at the end of filling the tub. This replaces Epsom Salts or any other remedy.

- *Allergies, cold, snoring:* Use as a nasal wash when sinuses are plugged. Due to the reduction in inflammation of nasal passages this technique can also reduce snoring!

- *Bug spray repellant, sunburn pain, bug bites, swelling:* Spray or soak a towel in 11.5 pH Ionized water and dab over the affected area for a minimum of 30 minutes twice a day.

- *Heartburn, indigestion, food poisoning, stomach flu:* Drink ¼ cup of fresh 11.5 Ionized Water, immediately followed by 25 ounces of 9.5 then do not eat or drink anything for 45 minutes. Repeat the next day only if necessary.

- *Arthritis, gout, muscle soreness, tissue injuries:* Since high alkalinity draws out acids, you can utilize 11.5 to soak in to "pull out" acids associated with inflammation, injury and pain.

- *Hangovers and migraines*: When you have a hangover or as soon as feel migraine coming on, drink several glasses of 11.5 pH Ionized Water.

- *Chemotherapy:* Drinking Ionized Water during chemotherapy can lessen the side effects and reverse metabolic acidosis. The antioxidants are good for any point during chemo. Apply 11.5 Ionized Water on the skin twice a day to heal chemotherapy burns.

- *Stroke*: If you feel a stroke coming, drink as much 11.5 Ionized Water as you can to provide your body with enough alkalinity to combat the acidosis causing the stroke.

- *Fruits & veggies*: Soak fruits and vegetables in 11.5 Ionized Water for a minimum of 5 minutes to clean off pesticides.

- *Ice cubes:* To help offset acidic drinks.

- *Rice, beans and legumes*: Soak for 5 to 10 minutes and rinse clean with low flow 9.5 Ph Ionized Water.

- *Meats:* Soak the meat in Ionized Water for 5 to 10 minutes for cleaning and tenderizing.

- *Laundry soap:* Use 1 to 2 quarts per load in place of laundry soap. Works beautifully for greasy smells like fast-food restaurant work clothes.

- *Stubborn stains on clothes, rugs and carpets*: Use as a degreaser for any type of cleaning. Clean oil stains by soaking the area in Ionized Water and letting it sit for 10 to 20 minutes. Then, blot out of carpets and wash off stain.

- *Clean oven, clogged sinks and tubs:* Replace your regular chemical laden cleaner with Ionized Water and a scratch pad.

- *Polish silver:* Soak and polish silver jewelry and silverware to restore lost shine.

- *Paint thinner:* After using oil base paints, use to clean up.

- *Get rid of gum and sticky stains*: Removes greasy, gooey, gummy, sticky problems.

2.5 Water – Strong Acid

Strong Acidic Water is a very potent anti-microbial. Its Oxidation Reduction Potential (ORP) is greater than +1100 mV. This makes it a great remedy for treating nail fungus, athlete's foot and other microbial infections. It also works as an effective mouthwash and to maintain excellent oral hygiene (make sure you rinse your mouth with neutral water after you gargle with strong acidic water). It can prevent infection when used to clean minor cuts and scrapes. It also helps stem bleeding and control pain. To add a toning effect to your bathwater, just add one or two quarts of Strong Acidic Water.

This water is widely used in hospitals and restaurants to reduce the spread of infection. It's used to disinfect medical equipment and instruments. It is also used to wash veggies and fruits, ridding them of bacteria and microbes, as well as to sterilise cutting boards and counter tops.

- *Anti-microbial:* Pre-rinse all fruits, vegetables and meat and let sit for one minute before soaking in 11.5 pH Ionized Water. This will kill all microbes and infections.

- *Disinfectant:* Use for disinfecting anything.

- *Antibacterial soap:* Use in place of anti-bacterial soap.

- *Hard water spots and rust:* Clean hard water spots off chrome and rust off of metal.

- *Facial lifting and tightening:* Spray face and neck (not eyes) and massage skin in upwards direction until dry. Finish with Beauty Water to tone skin.

- *Brush & Gargle:* Resolve periodontal disease and thrush, protect your teeth from root canals with strong acidic water. Wait one minute and then rinse with 9.5 water for 30 seconds to restore the natural pH.

- *Vomiting:* To stop vomiting and prevent nausea drink 2 to 3 tablespoons of 2.5 water.

- *Open wounds:* Cleans and disinfects open wounds, burns and infections. Stops wounds from bleeding and kills Candida, bacteria and pathogens. Use on cuts, scrapes to help stop bleeding. Clean twice daily until healed. Do not use any other ointments as they only attract microbes by keeping the area moist and sticky.

- *Infected sinuses:* Spray into nostrils twice a day for 2 days. Wait 2 minutes, then flush with 11.5.

- *Nail fungus:* Spray twice a day or soak in 2.5 Ionized Water.

- *Pink eye:* Spray this water into the infected eye several times throughout the day and the infection will clear up.

- *Sore throat, strep throat and cough:* Gargle 3-4 times per day or put in spray bottle and spray into throat.

- *Poison Ivy:* Spray on infected area as often as needed. It will soothe the itching and dry up the poison ivy much quicker.

- *Fever blisters and cancer sores:* Spray or gargle to stop and dry up both.

- *Moles and warts:* If you see something abnormal on your skin you may want to soak a gauze pad on a band aid with 2.5 Ionized Water and apply over the area. Change the dressing at least once a day. Often this process requires 30-60 days before seeing results.

Study by Dr. Ajay Sharma

B.Sc. (Agriculture); M.Sc. (Forestry), P.G. Dip. Ecology & Environment, Ph.D. (Plant Sciences) from Australia

An Introductory note about Anolyte

Anolyte is water with acidic pH (anions) prepared during electrolysis (ion exchange). Electrolysis is done in the presence of minor quantity of normal salt. Water collected at anode is called Anolyte or Electrolysed oxidised water (EOW) and it has positive ORP (ORP more than 400 mv kills bacteria, fungi and virus).

Anolyte is primarily a disinfectant that can restrict decay in fruits, food, meat, bread, sweets, crops, human body, water for several days.

Properties of Anolyte change with pH and ORP. Lower pH and higher ORP increases disinfection ability.

Anolyte is world's best sanitiser – liquid that kills bacteria as well as fungus, and restricts viruses. It kills bacteria in any medium (water, milk, juices, air) and from any surfaces (shelves, showcases, freezers, stores, taps, kitchen, fruits, vegetables, meat, chicken, poultry birds, domestic pets, human body, water tanks, water supply pipes, car interiors, hospital OPD areas, Operation Theatres). Practically, it kills bacteria and fungus from any surface, body, and medium. The uses are countless.

Major Applications of Anolyte

Homes (Domestic sector)/Schools/Mid-Day Meal Kitchens

1. Disinfection of drinking water; especially in areas using groundwater, and having septic tanks

2. Disinfection of salad, greens, vegetables, fruits and its shelf life prolongation;

3. Disinfection of fridges, Kitchen shelf, Utensils and Cooking plates

4. Improving shelf life of Achaar, Chutneys, fish, meat,

5. Disinfection of toddler kids (using diapers) – skin care

6. Disinfection of rooms where patients live, especially those suffering from asthma or infections

7. Disinfection of wounds, especially on pets, kids, and those of sugar patients

Safe for the Environment:

- The only elements introduced into the system are water, salt and electricity, and are all safe and environmentally friendly

- No disposal precautions

- Provides opportunity to reduce water usage

- Reduces volume of wastewater

- No adaptive resistance chance for microorganisms

- No environmental impact

- Fully biodegradable

- Satisfies the demand for implementing safer and more natural food products

Veterinary medicine

1. Treatment and preventive measures of diseases: wound sanitation of animals, watering of young animals (calves, piglets, nestling) at infectious intestinal diseases and for its prophylaxis, etc.;

2. Disinfection of incubatory eggs;

3. Sanitary treatment of animals cover-let including cow udder, etc.

4. Very useful for disinfection of patients having FMD, Mastitis, Diarrhoea, Intra Uterine infections

5. Anolyte or electrolysed fluid application by aerosol method for disinfection of cattle-breeding rooms at presence of people and animals;

6. Water treatment and disinfection at growing of young fishes and shrimps.

Water treatment and disinfection.

1. Water disinfection at water treatment plants;

2. Disinfection of water for drinking in flood hit areas (makes flood water bacteria free so that it may be used for drinking after filtration using clean cloth)

3. Disinfection of sewage, industrial wastewater, agricultural wastewater;

4. Disinfection of water in aquarium, terrariums;

5. Disinfection of pipes, cisterns and other tanks of any volume for supply or/and storage of drinking water.

Slaughter houses and Meat Shops

1. Disinfection of meat, fish and chicken against all pathogenic

2. microorganisms including Listeria

3. CIP-washing at slaughter houses to eliminate infection and stinking smell

4. Disinfection of carcasses of cattle and birds in slaughtering rooms and meat-processing plants;

Beauty Water 4pH – 6pH

It is beneficial for the overall health and beauty of your skin and hair. It tightens, softens, and provides an environment for healing excellent for many common skin conditions including:

Eczema

Psoriasis

Acne

Athlete's foot

Nail fungus

Insect bites/stings

Rashes (also relief from the itching of measles and chicken pox)

Sunburn

Cold sores

Diaper rash

Dry skin

Burns

Minor cuts and scrapes

Beauty water can be used as a spray or spritz throughout the day – the more often the better. When used instead of a conditioner after washing your hair, it tightens your scalp and it reduces tangles and brings out a radiant shine to your hair.

- *Facial Soap:* Clean face twice a day. Spray after cleaning.

- *Hair conditioner:* Spray hair after showering. Works as a conditioner, so you don't have to use an actual conditioner for your hair.

- *Facial toner:* Use as final rinse in shower or bath to tone and firm skin.

- *Rashes:* Spray on the affected area to soothe and heal rashes, including diaper rash.

- *Pets:* Bathe pets in Beauty Water for a more lustrous coat.

- *Plants:* Water indoor and outdoor plants for vigorous growth. This water can also revive dying plants.

- *Eggs & pasta:* Use to boil eggs and pasta.

- *Freezing food:* Spray on foods before freezing, especially for fish and shrimp so that foods do not lose their flavor.

- *Anthocyanins:* For washing and preparing fruits and vegetables containing anthocyanins: plums, grapes, cherries, strawberries, red cabbage, eggplant, soy beans, asparagus.

- *Fabric softener:* Use in rinse cycle during laundry. One gallon per load, for softer and fresher feeling fabrics.

- *Eyeglasses:* Clean lenses with ease. Simply spray on the water and wipe with a soft clean cloth.

- *Clean windows and mirrors:* Spray generously on the surface and wipe with a soft clean cloth.

- *Hardwood floors & ceramic tiles*: Polish and clean hardwood and ceramic floors without causing any damage.

- *Antiperspirant:* Assists the body's natural mechanism of getting rid of toxins, acts as a natural anti-perspirant and keeps the lymphatic system healthy.

8.5 Water to 9.5 Water

Consuming restructured, ionized water is the single most significant change you can make to support your body's natural environment. It's also the easiest way to do so.

- It will hydrate cells faster. This, in turn, will improve every function in your body. You'll notice better physiological function, softer skin, easier bowel movement and a spike in energy levels almost instantly.

- It will aid in the effective and efficient removal of toxins from the body. Think of it as your very own, personal filtration system that'll do all the detoxifying without making you do any extra hard work.

- Ionized water has a greater number of electrons, so it can neutralise free radicals, and thus alleviate pain, inflammation and many other symptoms.

- Ionized Water's alkaline pH levels can help maintain the balance of your body's default chemistry and keep your tissues from hardening due to acidosis.

– *Soups:* Cook all soups with 9.5 water for optimal boost of flavours and nutrients.

– *Stir fry:* Stir fry veggies with 9.5 water to steam to unlock all flavours and maintain nutritional balance.

– *Weight loss:* Drink before you eat a snack and or any meal. Wait for 30 minutes. Then eat if still hungry. Most folks are so dehydrated that their thirst mechanism is usually very weak. This makes them think that they are hungry instead of thirsty.

– *Grey hair:* This water can often restore the original hair color

– *Vision:* Rinse with this water daily to maintain healthy vision.

– *Spider veins:* Long term consumption of this water can promote cell repair and improve spider veins.

- *Aromatherapy:* Add any herb like rosemary or lavender to a spray bottle filled with Ionized Water. Leave it be for a couple of hours. Use as a spritzer in your house for a natural, effective and instant air freshener.

- **Cooking:** alkaline water can cut short cooking time by 25-30%. That's not all, steamed vegetables can retain more of their flavour, colour and nutrient profile when cooked with alkaline water. Beverages like tea and coffee lose their bitterness but retain their full flavour, when brewed with alkaline water.

Ionized Water and Your Kids

Can Children safety drink Ionized Water?

The answer is yes! Here are the industry wide recommendation:

- Baby food and formula should always be mixed with "Clean Water" i.e. 7.0 neutral pH.

- Medications: Alkaline water should not be given to children 30 minutes before or after receiving medications. It is recommended that you give them "Clean Water" or neutral water.

- Newborn to 1 years: Mix formula with "Clean Water".

- Children ages 1 to 5: You may give your child water with a pH of 8.5 to 9.0 to keep them hydrated. If the weather is hot be prepared to have it on hand. If your child is into sports or other strenuous activity, keep Ionized Water on hand as it will rehydrate them quickly.

- Children ages 6-12: You may give your child water with a pH of 8.5 to 9.0 to keep the hydrated. If the weather is hot be prepared to have it on hand. If your child is into sports or other strenuous activity increase the Ionized water up to 5 liters.

- Children 12-18 years: In this age range they can drink 9.0 to 9.5 Ionized Water to help keep them hydrated. This will also help with stamina, both physically and mentally.

- Always start your kids on Ionized Water 1.

Your Kids' Skin Cuts and Scrapes

Scraped, cut or torn skin is an inseparable part of every childhood, and quite simply unavoidable. Spray on the Strong Acidic water 2.5 pH at least once a day to promote healing. Once the cut has dried, soak a cloth or gauze pad in the Beauty Water and allow it to air dry. This tightens the skin and helps ease the pain.

Sore Throat and Strep Throat

How many times have your kids come home complaining of a sore throat? When they can't gargle, use a small spray bottle filled with Strong Acidic water and spray about 10 short pumps in their throat 3-4 times daily. It will not hurt them if they swallow some of it. The soreness goes away in minutes.

For teens and adults, gargle with the Strong Acidic Water 2.5 pH a few times daily works very well.

Diaper Rash: Spray Beauty water on skin after thorough cleansing to soothe and heal diaper rash.

Pink Eye: Spray 2.5 pH on infected eye several times throughout the day and it will clear up.

Ionized Water and Pregnancy

Ionized Water is safe and particularly useful during pregnancy. Here's a quick overview of some of its benefits:

- Have a healthier pregnancy

- Get rid of morning sickness!

- Increased lactation for moms who are breast feeding

- Ease discomforts of the last trimester

How Does Ionized Water Help Morning Sickness?

In the first few months of pregnancy a mother's body gives up alkaline minerals to the placenta. This happens to provide enough alkaline minerals to neutralize acidic discharges from the fetus for the next nine months.

This causes the blood pH to drop and become acidic, which causes morning sickness.

According to the doctors in Japan, this is why drinking alkaline water immediately relieves morning sickness. The mother should start drinking Ionized Water right since she and her partner start trying to conceive.

This will allow her body to build up its alkaline reserves. A healthy alkaline diet before pregnancy is also a wise choice. With adequate reserves built up the first few months the expecting mom can avoid mineral deficiency problems.

How Does Ionized Water Help Dehydration?

Pregnancy creates a very high demand for water in the body. As the fetus grows, over 1 trillion cell divisions take place. Each of them needs water! In order to stay hydrated during pregnancy, it's important to first eliminate dehydrating agents such as alcohol and caffeine.

Common Pregnancy Side Effects that can be remedied with Ionized Water:

Constipation: More than 50% expecting mothers experience constipation during pregnancy. Drinking Ionized Water rehydrates the body, lubricates the intestinal tract and helps flush out waste.

Swollen Hands and Feet: Did you know water retention is a sign of dehydration? This is common in the last trimester of pregnanc.

Lower Back Pain: Back pain is often a symptom of dehydration, because 67 the discs of the vertebrae are affected from the increased pressure of carrying the child. Alkaline Water can help lubricate the discs. Additionally, applying compresses of Ionized Alkaline Water to the lower back will provide pain relief.

Replenishing the body with Ionized Water rehydrates the body and provides relief from most symptoms.

Water ionizers make water safer because the machine's filters purify water before it is ionized. A water ionizer also allows you to control the strength of the Alkaline Water you want to consume. Alkaline Water made by a water ionizer contains essential minerals. For pregnant women, Alkaline Ionized Water provides Ionized Calcium, which is called "free Calcium" by doctors. Ionized Alkaline water helps replenish calcium loss in mothers-to – be and may strengthen the bones and teeth of the fetus.

Pregnant Moms and the 15 year Study

In a 15 year study, Japanese Water Institute and Kyowa Medical Clinic revealed these benefits of alkaline water for pregnant women.

- Smoother deliveries

- Reduced likelihood of jaundice

- Increased lactation

Alkaline water can help in smooth post-pregnancy recovery for new moms, as well as aid in lactation. Neutral water (7.0 PH) from a water ionizer is a safe choice for mixing formula and cereal for infants as well. The machine filters the water, but it is not ionized. Alkaline water is not for nursing infants. The mother's breast milk provides all the nutrition the baby needs.

Ionized Water and Your Oral Health

We have all had some form of dental problem in our lives, be it cavities, bleeding gums, toothaches, etc. One of the benefits I have discovered is how Alkaline Ionized Water has helped with my oral hygiene. Like many adults, I too face problems with my gums. Frequent cleanings help, but don't cure the buildup of tartar and gum inflammation. Using Kangen Water to brush your teeth, rinse your mouth, and scrape and clean your tongue can substantially improve overall oral health. It will also help heal canker sores and cold sores. Drink Ionized water to help keep that healthy balance.

Acidity and Oral Health

One of the sources of alkaline minerals is calcium from our teeth. Now, what happens when calcium is being leached from our teeth to ensure our body maintains a constant pH level?

Drinking Alkaline Ionized Water with a pH of at least 8.8 (Ionized Water of level 2 or 3) aids in killing the excess acids in your mouth that also contribute to tooth decay. With our diet, our mouth and saliva become more acidic than they should be.

Drinking Alkaline Water will handle this acidity so that it doesn't wreak havoc on your teeth.

Benefits of an alkaline rinse for gingivitis and periodontal disease:

It has been shown that rinsing with Acid (2.5 pH) Ionized Water kills the bacteria in your mouth that contribute to gum disease.

Dentists recommended using Alkaline Water (9.0 pH) to brush with and then rinsing your mouth with Acidic (2.5 pH) Ionized Water to kill any lingering bacteria. And a final rinse with Alkaline Water as you don't want to leave your mouth acidic. I also suggest rinsing your toothbrush with Acidic water to kill any bacteria on your toothbrush bristles.

This routine is safe for kids too…As it's an effective strategy to prevent cavities and fight bad breath!

What Your Dentist Knows About Your Health!

Who looks forward to a root canal or a procedure on the gums!?! Going to the dentist may not be the top activity for any of us, but you will be surprised just how much your dentist can tell about your overall health just by looking at your gums. For that reason alone, regular check-ups with your dentist is absolutely important and necessary.

Recent research shows there are 84% of patients with Coronary Artery disease also have periodontal disease. This not only raises the risk of heart disease, but hypertension and atherosclerosis. Evidence shows that oral infections are capable of raising inflammatory markers like C-reactive proteins and fibrinogen. These are red flags for atherosclerosis. It has been found that the patients with chronic Periodontitis have "Streptococcus mutans" – the same bacteria found in their dental plaque in the lining of the arteries of their hearts as well!

What can you do to keep your mouth and your heart healthy?

Clean your toothbrushes with 2.5 pH Acidic Water, to kill bacteria after each use. Brush using 9.0 to 11.5 Alkaline Water and rinse with 2.5 pH water for a minute and finish with a rinse with 9.0 pH Alkaline Water. Floss and use a water jet daily.

Regular brushing with the 2.5 pH acidic water helps whiten the teeth to give you that brighter smile. Change toothbrushes frequently or use a Sonic toothbrush daily. See your dentist regularly for checkups and cleaning.

For Digestive Health

Listed below are some of the common digestive problems and how Ionized Water can solve them.

ACID REFLUX

Taking acid blockers actually creates a vicious downward spiral when it comes to proper digestion. However, this issue is one of the fastest to respond to high quality ionized water. In natural medicine, we know that acid reflux is actually caused by inadequate production of hydrochloric acid at the proper time.

This is addressed quickly by following the Basic Water Protocol. Following the steps in this protocol actually re-trains the body to produce adequate amounts of hydrochloric acid at the right time. Supplementing your stomach acid with pH 2.5 strongly acidic water at the beginning of a meal assists your body in the proper breakdown of proteins.

CONSTIPATION

This problem is generally due to 2 major factors: food allergies and dehydration. However, in the case of ionized water there can be another important consideration . . . excess calcium in the water. When water goes through ionization it divides minerals according to charge, and calcium is diverted to alkaline water. In areas with hard water, it is necessary to pre-treat the water to remove the excess calcium. If this step is not taken, you may end up consuming too much ionic calcium in your drinking water, which can lead to constipation.

FOOD POISONING

The stomach is designed to completely irradiate foreign invaders like E. Coli, salmonella and other food borne pathogens. It does this by breaking their protein structures down with hydrochloric acid. So, the most effective way to assist your body in this destructive process is to drink cup of strongly acidic water every 2-3 hours until all symptoms have passed. It generally only takes 2 doses and you are back to normal! It is best to follow any bout of food poisoning with some probiotics and extra pancreatic enzymes for a few days. If you follow the Basic Water Protocol this regimen will encourage the good bowel bacteria to flourish.

IRRITABLE BOWEL SYNDROME

This problem is also associated with food allergies or sensitivities. However, in most cases this problem is actually exacerbated by a depletion of good bowel bacteria. This is generally the result of an upset in both the pH and hydration levels in the large and small intestines. The best place

to start is by following the Basic Water Protocol. This is the fastest way to reverse oxidation and return proper pH and hydration levels to cells.

Basic Protocol For Healthy Digestion

The ideal pH of the stomach is 1.8, which is very acidic. These acids help break down proteins in our food, so when stomach acids are inadequate, we must find a way to increase them. Once the food has been broken down in the stomach, it gets passed on to the small intestine, where the pH levels are an alkaline 8.5! This difference in pH is crucial to break down the fats in our foods to build healthy cells. The work of creating this shift in pH falls on the gallbladder which pushes enough amounts of bile in the small intestine and create an alkaline environment. The struggle begins when most people's gallbladders fail to produce sufficient bile, which is turn leads to indigestion and an imbalance in the intestine's pH levels. If the pH levels of the intestine aren't alkaline, the pancreas will not secrete the required enzymes for complete breakdown of food.

Dr. Parker mentions a 2-step approach to normalizing production of Hydrochloric Acid in the stomach.

- *First*, eliminate or neutralize it so that the system is forced to start production of the acid just in time for food to arrive! I used a little pH 11.5 ionized water to accomplish that task

- *Second*, add a bit more acid to the stomach to ensure the early stages of digestion are complete. To accomplish that I used just a little pH 2.5 water because it mimics a slightly weaker hydrochloric acid.

Dr. Parker further suggests another 2-step method to adjust the pH of the small intestine, to encourage the proper flow of bile and the release pancreatic enzymes to complete the breakdown of fats, proteins and carbohydrates during the final stages of digestion.

- *First*, eliminate the acids in the stomach. This helps alkaline drinking water to pass through the stomach directly into the small intestine without interference. Since water is the only substance we ingest that is not digested, it will pass virtually undisturbed through the stomach and into the small intestine where it is able to create an alkaline environment.

- **Second**, by consuming a large amount of alkaline-antioxidant rich drinking water at one time, there is an adequate amount of both water and antioxidants delivered that can be absorbed immediately into the blood stream and the lymphatic system. This encourages the body to eliminate toxins, which are responsible for creating an improper thickening of the bile, preventing it from freely flowing into the small intestine.

Following this protocol throughout the year will help create the ideal environment in your digestive system, encourage normal production and timely release of stomach acids and enzymes.

Chapter 7

Using Ionized Water to Aid Specific Problem

Important to Know: Cleansing Symptoms

People who are newly onto an Ionized Water regimen tend to experience what we call "Cleansing Symptoms". These include headaches, skin eruptions, dysentery, cough, etc. and are typically caused due the sudden release of stored toxins and acidic wastes from the body. If you're on an 8.5 pH water regimen, you should increase your water intake to help neutralize and flush the toxins out of your system. In case these symptoms persist, go back to the last pH water setting you were able to drink without experiencing any of these harsh symptoms. Keep increasing your water intake of this pH water until the symptoms stop. If the symptoms still persist, go back to drinking the water you're most comfortable with, but in increasing quantities until the symptoms subside.

45 minutes before your meals do the following:

- Drink ¼ cup pH 11.5 highly alkaline water

- Immediately followed by pH 9.5 drinking water – see formula below

- If you are taking any additional digestive enzymes, swallow them with this water

Now you MUST WAIT 45 MINUTES before you eat or drink anything else.

- With the first bites of your meal swallow 2-3 Tablespoons pH 2.5 strongly acidic water.

- If you are taking any vitamin supplements, swallow them with this water at the beginning of the meal.

- *Do not drink anything with your meal.* Anything you drink will be more alkaline than your stomach acids and will dilute them.

- Diluting your stomach acids dramatically impairs your ability to break down your food, especially proteins. Remember, if you are properly chewing your food you will not require any liquids to comfortably swallow your food. After a few weeks you will be out of the habit of drinking with your meals and will not even desire to do so.

For Healthy Skin

Skin is the largest organ of the body and its main function is to act as a barrier between the internal organs and the external environment. It plays a key role in blocking exposure to harmful pollutants, chemicals etc. While it excretes toxins, simultaneously it absorbs water and oxygen from the air to assist in hydration. The downside to this is that if you're not careful, anything you put on your skin will get absorbed into your bloodstream.

Use of harsh skincare products such as a bathing soap, along with exposure to harmful sun rays and pollution can break your skin's barrier and not only make it appear dull and aged, but also create an open pathway for harmful elements to enter your body. At the first sign of an assault of any harmful elements, your immune system sends out oxidizers that surround the microbes and destroy them. Once this is done, your body releases antioxidants to reverse the collateral damage. A part of this process involves increasing the skin's pH levels to mildly acidic to prevent any further bacterial growth.

Throughout this process, it's crucial to help restore and maintain your skin's pH balance, as that's the key to healthy, radiant skin.

Here's how Ionized Water can help you in your basic skincare:

- **Step 1** – Heavily mist skin with pH 11.5 highly alkaline water to remove excess oils and any makeup.

- **Step 2** – Cleanse your face with a non-soap cleanser and rinse thoroughly with pH 4-6 mildly acidic water, then pat dry with a soft cloth.

- **Step 3** – To tone your skin, lightly mist with pH 4-6 mildly acidic water stored in a glass bottle equipped with a fine spray mister and allow your skin to air dry.

- **Step 4** – Apply moisturizer.

- **Step 5** – Repeat morning and before bed.

Basic Water Protocol for Acne Prone Skin

Ionized Water can not only help you maintain a healthy skin, it can also diminish acne and keep it at bay. Follow this protocol:

- **Step 1** – Rinse your face with pH 11.5 highly alkaline water to remove excess oils.

- **Step 2** – Cleanse with a non-soap cleanser and rinse with pH 4-6 mildly acidic water. Pat your skin dry with a soft cloth.

- **Step 3** – With a cotton swab dab pH 2.5 strongly acidic water on any active blemishes or broken skin, leave on for at least 30 seconds to 1 minute. Gently pat dry.

- **Step 4** – Mist your face heavily with pH 11.5 water. Wait 1 minute and then pat dry.

- **Step 5** – Tone your skin with pH 4-6 mildly acidic water stored in a glass bottle equipped with a fine spray mister.

- **Step 6** – Repeat morning and before bed.

- **Step 7** – If you live in a dry area, spritz your face with mildly acidic water of pH 4-6 multiple times a day, to keep skin hydrated.

Basic Water Protocol for Blisters and Burns

Soothe blisters and burns and speed up the healing process with Ionized Water. Here's how –

- **Step 1** – Because blistered skin is very susceptible to secondary staph infections, it is important to avoid any skin-to-skin contact. Make sure you sterilize the affected area and anything that comes in contact with your skin, with acidic water of pH 2.5.

- **Step 2** – Soak affected area with a sterile cotton pad saturated with highly alkaline water of pH 11.5 to remove excess oils or dried fluids from the skin.

- **Step 3** – Cleanse your skin by spraying on a mixture of a very mild non – soap cleanser diluted in the ratio 10:1 with pH 4-6 mildly acidic water.

- **Step 4** – Rinse the area thoroughly with mildly acidic water of pH 4-6.

- **Step 5** – Mist the affected area generously with strongly acidic water of pH 2.5. Leave it be for 1 minute and then gently pat it dry with a sterile cotton pad.

- **Step 6** – Heavily mist the area with alkaline/antioxidant water of pH 11.5. Allow your skin to air dry or gently pat the area with sterile cotton pads.

- **Step 7** – Thoroughly mist with mildly acidic water of pH 4-6 stored in a glass bottle and equipped with a fine spray mister.

- **Step 8** – Repeat the above steps in the morning and before bedtime while blisters are intact. Once the skin is exposed it may be necessary to repeat Steps 2, 4 and 5 up to 6 times per day.

- **Step 9** – Mist lightly with mildly acidic water of pH 4-6, 2-4 times per day to keep the areas hydrated but make sure the skin is not moist.

NOTE: Should debridement become necessary on severely burned or abraded tissue, use pH 11.5 water to loosen the top layer of tissue, once removed, flood with pH 2.5 acidic water to clean and disinfect, follow with another rinse of pH 11.5 water to flood the affected areas with antioxidants, end with a final rinse of pH 4-6 water to balance the pH of the tissue.

Follow up with regular care as outlined above.

Basic Water Protocol for Cuts, Scrapes and Wounds

Cuts, scrapes and open wounds acts as doorways to a host of bacteria, viruses and infections. Keep it clean, disinfected and aid in the healing process with Ionized Water in the following way.

- **Step 1** – Wear non-latex gloves and sterilize anything that can come incontact with the affected area, with acidic water of pH 2.5.

- **Step 2** – Clean the affected area with a sterile cotton pad soakedin highly alkaline water of pH 11.5, to remove excess oils or dried fluids from the skin – it also removes residue from tape or adhesive bandages.

- **Step 3** – Cleanse by soaking or flooding the area with a mixture of 1 part very mild non-soap cleanser diluted with 10 parts mildly acidic water of pH 4-6.

- **Step 4** – Rinse thoroughly with pH 4-6 mildly acidicwater.

- **Step 5** – Heavily mist or flood the wound, cut or scrape and surrounding area with pH 2.5 strongly acidic water. Leave it be for 1 minute, and then gently pat dry with a cotton pad.

- **Step 6** – Heavily mist or flood the area with pH 11.5 alkaline/ antioxidant water. Let your skin air dry or gently pat it dry with sterile cotton pads.

- **Step 7** – Thoroughly mist with pH 4-6 mildly acidic water stored in a glass bottle and equipped with a fine spray mister.

- **Step 8** –Dress with gauze pad or bandage.

- **Step 9** – Repeat morning and before bedtime for scrapes and minor cuts. In case of deeper cuts or infections it may be necessary to repeat Steps 2, 4 and 5 up to 6 times per day.

- **Step 10** – Mist lightly with pH 4-6 mildly acidic water 2-4 times per day to keep the area hydrated but not moist.

Basic Water Protocol for Eczema and Psoriasis

Psoriasis is a chronic skin disorder commonly characterized by inflamed red patches covered by a silvery, flaky layer of skin that appears all over the body, including the limbs and scalp.

Psoriasis, as a condition, mainly occurs when the immune system mistakes normal functioning skin cells for pathogens, which in turn, forces the immune system to send inaccurate signals to the brain. This reaction causes the skin to overproduce new skin cells.

When this happens, the surface layer of the skin starts taking on an inflamed appearance. There are five types of psoriasis; plaque psoriasis is considered most common. Psoriasis commonly affects the skin and the elbows of the knees, in addition to the scalp, hands and feet.

Psoriasis is one of many immune-related skin conditions. It's a disease affecting the skin on a lifelong basis. There's also no cure, as of this time. The condition itself is known to follow a pattern of acutely occurring flare – ups, alternating with periods of remission.

Psoriasis, however, can be controlled through various natural treatments, mainly for the purpose of subsiding the oft uncomfortable symptoms. All – natural treatments are an alternative to regular psoriasis medication and topical creams. These treatments provide a way to relieve symptoms without the harmful effects of medications and topical creams:

Apple cider vinegar washes. Apple Cider vinegar's alkaline properties help reduce the reddening inflammation of psoriasis. You can try taking baths enhanced with 2 cups of apple cider vinegar at least three times a day before showering off. It's also recommended to drink a glass of water containing 1 tablespoon of apple cider vinegar to help hydrate and soothe your skin from the inside.

Oatmeal. Oatmeal contains oat extract, which has a soothing effect on skin. Oatmeal provides a short-term remedy for psoriasis, as it helps soothe inflamed skin, but not reduce its appearance.

Fish oils. Fish oil contains a high ratio of healthy Omega-3 fatty acids. These fatty acids have anti-inflammatory effects on the body, which helps regulate the immune system, which in turn, may help reduce the chance of experiencing psoriasis flare-ups. Take two capsules of fish oil, twice a day with meals.

Skin-friendly foods. Dark chocolate, red and blue berries and even coffee are some of the foods containing anti-oxidant-friendly polyphenols, which force cell death in cancerous cells. Onions and green tea also harbor helpful dietary flavonoids called quercetin that help regulate the immune system.

Exercise. Even exercise helps regulate psoriasis flare-ups. Take a walk or jog for at least 45 minutes per day if you can. This helps promote blood circulation throughout the body and help blood flow improve in your skin.

Ionized Water and Psoriasis

See how pH 2.5 Acidic Water from our Ionized Water Ionizers can be used to treat various psoriasis conditions.

- **Step 1** – Rinse your skin with highly alkaline water of pH 11.5 to remove excess oils from the skin. Take time to soak areas that have formed a clear "crust". Dab the area until it's clear, but take precautions to not rub it.

- **Step 2** – Cleanse with a very mild non-soap cleanser and mildly acidic water of pH 4-6.

- **Step 3** – Rinse thoroughly with mildly acidic water of pH 4-6. Once done, gently pat the area dry with a lint free towel.

- **Step 4** – Saturate a cotton ball with pH 2.5 strongly acidic water and generously apply on wherever there is broken skin. This step is essential to reduce the chance of contracting a secondary staph infection. After 1 minute, pat the area dry very gently with a lint free towel.

- **Step 5** – Heavily mist or rinse affected area with alkaline/antioxidant water of pH 11.5. Allow the area to air dry or pat dry with a clean lint free towel after 2 minutes.

- **Step 6** – Tone the skin with mildly acidic water of pH 4-6, stored in a glass bottle that is fitted with a fine spray mister. Allow your skin to dry up naturally.

- **Step 7** – In mild cases, repeat morning and bedtime, for more severely broken skin repeat 3-4 times per day.

- **Step 8** – Mist with pH 4-6 mildly acidic water to keep skin well hydrated every 1-2 hours or as needed to control any discomfort. Air dry completely between hydration treatments.

Basic Water Protocol for Herpes Type 1 and 2

The best treatment for oral or genital herpes is immediate treatment at the very first sign of an outbreak. Speedy treatment can actually stop an outbreak before it erupts. Follow the Basic Water Protocol to prevent an outbreak.

- **Step 1** – Early treatment starts with pH 2.5 strongly acidic water. Soak a cotton swab or cotton pad and apply to the skin anywhere you have that "tingling" sensation. Leave on skin for at least 1 minute.

- **Step 2** – Follow with pH 11.5 strongly alkaline water to remove oxidative damage to the affected areas. Leave on skin for at least 1 minute or until it's dry.

- **Step 3** – Finish by misting the area with pH 5.5 mildly acidic water and let air dry.

- **Step 4** – Repeat this hourly during the first day of symptoms.

- **Step 5** – If symptoms persist or an eruption occurs, repeat this process 6-8 times per day.

Basic Water Protocol for Rashes & Contact Dermatitis

Rashes and contact dermatitis cause extreme discomfort to those affected. Even unintentional scratching can open the door to unwanted infections. So, treating rashes at the earliest is a great way to block the rash from spreading. However, rashes can appear rapidly leaving treatment to reduce red, itchy skin as your only option.

- **Step 1** – The first step is generally to remove any potential allergens that could be causing the rash or contact dermatitis; changing the pH and temperature of the skin in case of heat rash. Thoroughly cleanse area with a solution of 1-part mild non-soap and 10 parts pH 11.5 highly alkaline water.

- **Step 2** – Rinse thoroughly with pH 4-6 mildly acidic water.

- **Step 3** – If any lesions or open skin appears, follow up with pH 2.5 water applied with a spray bottle or a saturated cotton pad.

- **Step 4** – Follow by flooding area with pH 11.5 highly alkaline water then pat dry.

- **Step 5** – Finish by misting the affected area with pH 5.5 mildly acidic water.

- **Step 6** – Repeat 2-6 times per day and mist with pH 5.5 mildly acidic water as often as necessary to relieve the discomfort.

Basic Water Protocol for Rosacea

Rosacea is characterized by excessive reddening of the skin, especially on the cheeks, chin and nose. Although it's not commonly diagnosed as such, rosacea is actually an inflammation of the capillaries and the skin tissue itself. This makes it an excellent candidate for treatment with antioxidant rich, ionized water!

Capillaries are often the first to suffer in this skin condition. The best place to begin the healing is with a big dose of antioxidants and hydration.

- **Step 1** – Gently wipe the area with highly alkaline water of pH 11.5 with a cotton pad to remove excess oils.

- **Step 2** – Cleanse with non-soap cleanser and rinse with pH 4-6 mildly acidic water. Pat dry with a soft cloth.

- **Step 3** – Mist the skin lightly with pH 2.5 strongly acidic water. Allow your skin to air dry or gently pat dry. This step destroys the bacteria that accompany the inflammation.

- **Step 4** – Mist heavily with pH 11.5 water. Wait 1 minute, and then pat dry. Fresh 11.5 water contains many antioxidants that work to reverse the inflammation and its devastating results. In just a few days some of the puffiness will disappear.

- **Step 5** – Tone your skin with pH 4-6 mildly acidic water stored in a glass bottle equipped with a fine spray mister.

- **Step 6** – Repeat the above steps in the morning and before bed.

- **Step 7** – Moisturize as usual. Make sure your face creams contain no parabens, propylene glycol, alcohol, synthetic fragrances or petroleum products.

- **Step 8** – Lightly mist the face with a fine spray of pH 4-6 mildly acidic water multiple times per day to keep the skin hydrated.

Basic Water Protocol for Shingles

Shingles happen when the immune system wasn't able to completely flush out the chickenpox virus. Some remnants lay dormant in the nervous system, slowly gaining strength over the years.

One day, in a trauma to the nervous system, your body can no longer keep this virus dormant. When that happens, it's as though the fiery inferno was unleashed! Searing, mind-numbing pain begins, followed by an angry red rash and eventually small pustules appear. These symptoms are often severe and last for 4-6 weeks!

One of your best insurance policies against an outbreak of Shingles is a healthy immune system. That starts with the Basic Water Protocol.

- **Step 1** – At the first signs of tingling or pain, mist thoroughly with pH 2.5 strongly acidic water and allow it to air dry. Because the skin is porous it will absorb some of the oxidizing agents which can help to destroy the virus.

- **Step 2** – Follow with pH 11.5 strongly alkaline water. Heavily mist the affected area, and allow to air dry.

- **Step 3** – Finish treatment with a heavy mist of pH 4-6 mildly acidic water.

- **Step 4** – Mist affected area with pH 4-6 mildly acidic water as needed to reduce the pain.

In case rashes or lesions have already made an appearance:

- **Step 1** – Heavily mist the affected area with pH 11.5 strongly alkaline water to remove any dried fluids or oils.

- **Step 2** – Cleanse the area with a solution made from 1 part very mild, non-soap cleanser and 20 parts pH 11.5 strongly alkaline water.

- **Step 3** – Rinse thoroughly with pH 4-6 mildly acidic water and pat or air dry.

- **Step 4** – Mist thoroughly with pH 2.5 strongly acidic water and allow it to air dry.

- **Step 5** – Heavily mist the affected area with pH 11.5 strongly alkaline water and air dry.

- **Step 6** – Next, heavily mist the area with pH 4-6 mildly acidic water. Allow your skin to air dry or gently pat dry with a soft, lint-free cloth.

- **Step 7** – Mist the area with pH 4-6 mildly acidic water as needed to reduce pain.

For Healthy Hair

The health of your hair is the first indication of your overall health. And beautiful hair begins with a healthy and hydrated scalp.

The onslaught of shampoos, conditioners and styling products loaded with sulphates, parabens, synthetic fragrances and petroleum products robs the hair and scalp of its moisture and balance of natural oils. This, in turn, results in dry, brittle and dull hair and a flaky scalp.

However, with a routine revolving around regular use of Ionized water, this damage can be rectified. Here's how.

- **Step 1** – The easiest way to effectively shampoo your hair is to actually wash it in the sink. When washing your hair in the shower the majority of the shampoo is applied to the top of your head, effectively drying out the scalp in an area that receives the harshest treatment from products, styling tools and the sun.

- **Step 2** – Mix desired amount of shampoo with 1 cup warm water to thin it.

- **Step 3** – Adjust the water temperature to lukewarm and thoroughly wet your hair. The water is a perfect temperature when it's warm but not so warm that the heat sensor in your ionizer gives out an alarm.

- **Step 4** – Apply water/shampoo mixture evenly over entire scalp and saturate hair. Massage your scalp and work the shampoo mixture through hair.

- **Step 5** – Rinse hair and scalp thoroughly with pH 4-6 mildly acidic water.

- **Step 6** – Do NOT condition your hair unless it is exceptionally dry or coarse. Conditioning average hair will make it rather limp and difficult to style.

- **Step 7** – Style as usual.

Ionized Water to Keep Your Hair Stylish For style touch-ups

For style touch ups between shampoos, spritz hair lightly with pH 4-6 mildly acidic water before using a flat iron or hair drier. This will revitalize and hydrate both curly and straight hair.

Tame the frizz

If your hair tends to be frizzy, mix a few drops of natural conditioner or gel mixed with mildly acidic water to instantly tame frizzy hair.

Heat styling

For those who frequently use a flat iron, curling iron or hot rollers, mix a few drops of conditioning serum or styling gel with pH 4-6 mildly acidic water. Spritz this mixture lightly onto your hair before using styling tools. This helps to protect the hair from the drying effects of heat.

Ionized Water for Hair and Scalp Problems

Alopecia & baldness

These conditions are often the result of an imbalance in hormones, pH levels of the scalp or fungal infections. To combat this problem, rinse your scalp with mildly acidic water of pH 4-6. Follow this up by spritzing your scalp with mildly acidic water at least twice a day. The treatment may take a few months to show measurable results.

Dandruff & itchy scalp

The scalp is particularly prone to dryness from soapy residue, excessive styling products and hair driers. They can lead to pH changes, clogged pores and dry, itchy and scaly skin. Mildly acidic water is the perfect solution to all these conditions.

For Healthy Eyes

Overexposure to harmful chemicals, smoke as well as electronic devices have left our eyes more tired and strained than ever, making it prone to infections as well as poor eyesight.

Here's how you can use Ionized Water to clean your eyes and keep it away from infections.

- **Step 1** – You will need an eye cup. Upon removing the eye cup from its packaging, soak in pH 2.5 strongly acidic water for 1-2 minutes to clean and disinfect.

- **Step 2** – Rinse the eye cup thoroughly with pH 11.5 highly alkaline water.

- **Step 3** – Fill the cup, following package directions, with pH 11.5 highly alkaline water.

- **Step 4** – Place cup firmly around one eye, keeping your eye open, tilt your head back and gently roll your eye as though you were attempting to look up, down, and from side to side. Continue this for approximately 1 minute.

- **Step 5** – Now that you have completed one eye, toss the water away.

- **Step 6** – Rinse the eye cup thoroughly with pH 2.5 strongly acidic water (Repeat steps 2-7 for your other eye).

Frequency: To maintain healthy eyes follow this protocol 1-3 times per week.

When working to improve any eye condition, follow this protocol at least 2 times daily and up to 10 times per day.

In the case of an infection, saturate a sterile cotton pad with pH 2.5 strongly acidic water and rest on your eye – with your lids closed – for 1 minute. Then follow the entire protocol as outlined above.

Caution: Never use strongly acidic water in your eyes! The natural pH of your tears is quite alkaline, so simply give your body a bit more of what it is designed to use.

For Healthy Ears

The eardrum is quite sensitive and regular earwax removal and cleaning techniques can prove to be rather fatal. Follow the Basic Water Protocol for cleaning your ears with Ionized Water in the best way possible.

Directions

- **Step 1** – Sterilize a small dropper bottle with strongly acidic water (pH 2.5)

- **Step 2** – Rinse thoroughly with highly alkaline water (pH 11.5)

- **Step 3** – Fill bottle, tightly cap and set in a bowl of very warm water (not boiling) about 5 minutes

- **Step 4** – Test the temperature of the water by placing a drop on the inside of your wrist (like checking the temperature of baby formula) – this is very important since temperatures a few degrees cooler or warmer can create dizziness or vertigo!

- **Step 5** – Tilt your head as far to the side as possible

- **Step 6** – Place only 1-2 drops in your ear

- **Step 7** – Place cotton in the ear and leave for 1-5 minutes

- **Step 8** – Repeat steps 5-7 in the other ear

Frequency:

To maintain healthy ears, follow this protocol no more than 1 time per month. Earwax provides a natural protective barrier for your ears; excessive cleaning can upset the normal balance in the ear canal. Only clean using an irritating syringe if you have a waxy build up that impairs your hearing

General Information

Earwax is a fatty substance. Chewing is the mechanism by which it naturally moves along the ear canal. Excessive stress and anxiety can cause excessive earwax production.

NOTE: *Do not use strongly acidic water or small objects to clean your ears!*

Ionized Water for Nasal Lavage

Since your nasal passages are one of the first lines of defense your body uses to stop potential invaders, keeping the mucosal lining healthy and vital will go a long way toward keeping you healthy.

Supplies

- Neti Pot

- pH 11.5 highly alkaline water for daily use

- pH 2.5 strongly acidic water for use during active infection

Directions for daily use

- **Step 1** – Adjust your water temperature so that it is approximately the same temperature as your skin.

- **Step 2** – Fill your Neti Point with warm pH 11.5 highly alkaline water. It is unnecessary to use additional herbs in your nasal lavage.

- **Step 3** – Follow the directions for use that came with your Netti Pot.

- **Step 4** – Repeat with the opposite nostril.

Using a Neti Pot with highly alkaline water on a daily basis dramatically improves symptoms that accompany allergies – both seasonal and chronic. Using pH 11.5 highly alkaline water provides tons of antioxidants to reduce inflammation and irritation, as well as rebalances the mucosal lining of your sinus passages. In both occasional seasonal allergies and chronic allergies keeping the sinus passages properly moist and free of debris offers a great deal of relief. This technique has been used for hundreds of years in the Ayurvedic tradition of medicine.

For Oral Health

Optimal oral health is essential for your overall well-being. Infections or diseases in the oral cavity can spread to the nervous system causing permanent damage or even proving to be fatal. Fortify your dental hygiene practices with Ionized Water.

- **Step 1** – Rinse with pH 2.5 strongly acidic water, scrape your tongue, rinse, scrape again.

- **Step 2** – Brush teeth with pH 2.5 strongly acidic water, repeatedly dipping your brush in the water.

- **Step 3** – Rinse again with pH 2.5 strongly acidic water.

- **Step 4** – Floss teeth then rinse with pH 2.5 strongly acidic water.

- **Step 5** – Finally, rinse with pH 9.5 drinking water.

Frequency

Follow this protocol 1-2 times daily, in the morning and before bed. But it's mandatory that you follow this routine before bedtime at least.

For Colon Hydrotherapy

Equipment & supplies

- Gravity fed colonic irrigation equipment

- 9.5 Alkaline Water slightly warmer than room temperature

Directions

- **Step 1** – Adjust the temperature of your supply water to be slightly warmer than room temperature.

- **Step 2** – Fill your gravity fed tank with pH 9.5 alkaline water

- **Step 3** – Continue with your normal fill, release cycles

- **Step 4** – A final retention fill of pH 11.5 highly alkaline water can be used in cases of chronic constipation or following a round of antibiotics

Frequency: Colon hydrotherapy must be considered as an occasional cleansing routine. Best results are generally obtained when done 3 days in a row.

CAUTION – Over use of colon hydrotherapy creates an imbalance in the intestinal flora. Imbalance in intestinal flora leads to chronic, systemic depletion of vitamin B as well as inadequate breakdown of insoluble fibers necessary for beneficial bowel bacteria to grow and flourish.

For Yeast Infections & Vaginal Dryness

Supplies

Standard Douche/Enema Bag available at any drug store pH 4.5 Acid Water

Direction:

- **Step 1** – Adjust water temperature to match skin temperature, then select the 9.5 setting

- **Step 2** – Fill your bag with approximately 2 cups of the water from the gray hose – this will be approximately pH 4.5

- **Step 3** – Irrigate according to package insert directions

Frequency: In case of Vaginal Yeast Infections – 1 to 2 times per day for 2 day s Vaginal Dryness – once week for 1 month should resolve your problem if you are consuming enough water according to the Basic Water Protocol

***Caution** – Do not douche regularly! Douching upsets the normal pH balance of the vagina as well as the normal colonization of good bacteria. Upsetting this balance and can lead to infection, excessive discharge, overgrowth of candida and E. coli, all of which can lead to cancer.*

For Renal Disorders

From urinary tract infections to kidney disease and dialysis, water is a major topic when it comes to the hard-working kidneys! Here are a few suggestions when it comes to drinking ionized water and renal disorders.

Basic Water Protocol for Urinary Tract Infection

Most UTI's begin in the bladder before making the journey to the kidneys. Beginning the Basic Water Protocol at the first signs of discomfort is the best way to treat a UTI. Generally, a UTI starts due to a change in the pH of the bladder. This allows bacteria to adhere to the sides of the bladder wall and begin to multiply. Ideally the pH of the first morning's urine should fall between 5.8 and 7. A pH reading below 5.5 is the ideal breeding ground for bacteria in the bladder. A pH above 7.5 is a clear signal of infection and is responsible for the stinging or burning that often accompanies a UTI.

NOTE – At the first sign of an infection follow begin this routine. However, if your symptoms persist for more than 2-3 days, or worsen with treatment, consult your physician.

UTI Water Protocol based on body weight

Drink 120% of your body weight in ounces of water.

Based on the calculations for the Basic Water Protocol, consume the amount based on .75 oz before each meal.

The remainder of the water should be consumed at a rate of one 8-12-ounce glass every 1 to 1.5 waking hours.

Add 1 tablespoon *unsweetened* organic cranberry juice concentrate to 6 of these glasses of water. Cranberries lower the pH of the urine and contain compounds that keep bacteria from sticking to the walls of the bladder.

For Healthy Nails

Nail fungus is a systemic problem and must be treated internally as well as externally. Fungal conditions can be the result of damage to the nail or can be transferred from improperly sanitized manicure or pedicure tools.

To guard your nails against infection, take a small spray misting bottle of pH 2.5 strongly acidic water with you to your nail salon. Ask them to spray their tools before use and your nails before polishing.

- **Step 1** – Every morning and every night before bed spray the nails, the cuticles and the surrounding tissue with pH 2.5 strongly acidic water and allow it to air dry.

- **Step 2** – Depending on the severity of the fungal infection you will need to soak the nails in a basin of warm pH 11.5 highly alkaline water for 20 minutes 2-4 times per week.

- **Step 3** – After soaking, pat your feet dry or allow it to air dry completely. Then, spray nails with pH 2.5 strongly acidic water, and allow your feet to air dry again.

- **Step 4** – Apply a natural antifungal like Tea Tree, Peppermint or Lavender Essential Oil as a moisturizer.

TIP – Soak socks and cotton glove liners in pH 2.5 strongly acidic water 10 minutes or as long as overnight before putting them into laundry, to kill microbes. Until fungal nails are completely resolved do not wear socks or glove liners more than one time between laundry.

Ionized Infants and Children

It is a myth that babies and children are already alkaline so ionized water is either dangerous or unnecessary! If this were true then they would not be sick! Also remember the most important aspect of ionized water is its antioxidant property. We are all exposed to oxidation from the moment of conception.

Their smaller bodies are exposed to the same environmental and internal oxidative stressors as adult bodies, they simply have not had as much exposure time.

Use a similar formula for determining the amount of water to give that we use for adults since amounts are always weight dependent. Because children are generally very active and their bodies are growing rapidly, they can actually consume a slightly higher percentage of water per pound of body weight.

- Using a calculator, multiply your child's body weight by .65 and by .85.

- The first number represents the minimum number of ounces a child's body needs *just to reach a state of hydration.* The second number represents the maximum number of ounces ideal for more therapeutic results.

- Now divide that number by 5. This number represents the ideal amount of pH 9.5 ionized drinking water to consume in one sitting – within 10 minutes to deliver greater numbers of antioxidants.

Baby Formula

Baby formula is an unnatural food for babies. When formula is the only option available, mix it with pH 9.5 drinking water. This helps break down some of the essential fatty acids, making them easier to assimilate. Fats are crucial to the proper development of the brain, the nervous system and cell membranes.

Skin Care for children

Skin irritations, rashes, cuts, scrapes and even diaper rash are all great candidates for treatment with various pH levels of ionized water.

Remember it's never too early to start skincare! Just like adults, your child's skin should be clean and hydrated. Follow the Basic Skin Care Protocol and make a game out of the misting – kids love it!

Protocol for oral health for children

Beginning with your baby's first teeth, oral health is essential to avoid cavities.

- **Step 1** – Saturate an infant toothbrush or a nubby wash cloth with pH 2.5 strongly acidic water and thoroughly clean teeth and gums.

- **Step 2** – Once your child has more teeth, begin gentle flossing. Soak floss in pH 2.5 strongly acidic water before using.

- **Step 3** – Saturate the toothbrush or cloth with pH 11.5 strongly alkaline water and clean teeth and gums.

- **Step 4** – Gently irrigate teeth and gums with a bulb syringe filled with pH 9.5 drinking water to rebalance the pH of the mouth.

Ionized Water for your Pets

Directions:

Fill each of 3 bowls with pH 8.5, pH 9.0 and pH 9.5 drinking water respectively. Places these bowls in the spots where your pet normally goes to drink water. Wait and watch which pH water your pet prefers, and use that one as its drinking water.

Digestive and skin disorders

Many animals suffering with digestive disorders, skin disorders, parasitic infections and arthritis have been helped dramatically when their water was switched to high quality ionized water.

Skin infections, rashes and itchy skin should be treated in the same manner as human skin.

If you find your pet excessively licking, chewing on their paws or picking out feathers, it may be a sign of allergies. Since their hair, fur or feathers are fantastic places for pollen to collect, your pet is then exposed to more allergens during the course of their natural self-grooming. In these instances, either mist and dry or simply wipe your pet down with a cloth that has been dampened with pH 5.5 mildly acidic water 2-3 times a day until their excessive behavior subsides. As a part of your regular routine, wiping the coat down 1-2 times per week is a great way to prevent allergens from becoming a problem.

Basic Water Protocol for bathing pets

- **Step 1** – Prepare to bathe your pet by collecting pH 4-6 water and storing it in gallon jars until you have enough water or make fresh mildly acidic water. Warm the water slightly before bath.

- **Step 2** – Dilute the shampoo with water to make it easier to wash your pet entirely with little shampoo. Mix 1 teaspoon shampoo to 1 cup pH 11.5 water. This is enough for a very furry medium sized dog.

- **Step 3** – Saturate the entire coat with diluted shampoo mix and lather well.

- **Step 4** – Rinse thoroughly with pH 4-6 mildly acidic water.

- **Step 5** – For long haired cats and dogs, conditioning may be necessary. Combine 1 teaspoon conditioner with 1 cup pH 4-6 mildly acidic water (enough for a small dog) and saturate hair, massaging into coat.

- **Step 6** – Rinse thoroughly with pH 4-6 mildly acidic water.

- **Step 7** – Towel dry then blow dry on cool setting. Never use hot air to dry your animals. It is too drying to their skin. Always avoid their ears, eyes and nose. Animals are much more sensitive to drying air and the noise of the hair dryers.

- **Step 8** – Never brush or comb a wet animal. Allow the coat to naturally air dry or dry with a blow dryer until just damp. Brushing or combing while wet will stretch and break the hair. This is more important to keep in mind for long haired animals.

- **Step 9** – For regular brushing and grooming, fill a spray misting bottle with pH 4-6 mildly acidic water and use it to dampen their coats and brush as usual.

Basic Water Protocol for tear stains

- These darkened stains on light colored coats are both unsightly and difficult to manage. It is often a sign of a pH imbalance. So, the logical place to begin is with your pet's drinking water!

- The next step would be to thoroughly bathe your pet.

- Before preparing the pH 4-6 Mild Acid Water for the bath, pre-treat the stained areas with pH 11.5 highly alkaline water.

- After bathing and brushing your pet, continue to clean the affected area at least once a day with a soft cloth dipped in highly alkaline water. This will eventually fade the tear stains and prevent them from becoming a problem in the future.

Basic Water Protocol for Oral Healthcare

While cats tend to have fewer problems with tooth decay and gum disease, dogs require regular dental care.

Supplies

Toothbrush – you can either use a regular adult toothbrush for medium to large dogs, a child's toothbrush for small dogs or a special brush that fits on your finger. These are especially helpful for dogs that are hesitant to let you brush their teeth.

pH 2.5 strongly acidic water

pH 11.5 highly alkaline water

pH 9.5 drinking water

Bulb syringe for small to medium dogs and a turkey baster for large and extra-large dogs

Directions

- **Step 1** – Fill bulb syringe with pH 2.5 strongly acidic water and irrigate along the entire gum line.

- **Step 2** – Saturate toothbrush in pH 2.5 strongly acidic water. Begin brushing on the outside of the teeth. Dip brush into the water frequently. Move to the inside of the mouth, taking time to brush thoroughly. If your pet isn't used to regular brushing, you may have to start with only 1 quadrant a day and work your way to a full mouth brushing.

- **Step 3** – Fill bulb syringe with pH 11.5 highly alkaline water and irrigate along the entire gum line as well as between teeth.

- **Step 4** – Fill bulb syringe with pH 9.5 drinking water and irrigate mouth to restore natural pH.

Frequency

If your pet has gum disease or tooth decay, it is imperative that you brush their teeth every day until the problem resolves. This could take several weeks or even months depending upon their overall health and seriousness of disease.

For regular maintenance of healthy teeth and gums, brushing 1-2 times per week is adequate.

NOTE – If teeth have significant tarter build up, it's a great idea to make an appointment for a thorough cleaning. Small breeds tend to be at a greater risk of gum disease and tooth loss, so they should have their teeth brushed more often than medium or large dogs. Cats generally only need to have their teeth brushed as they age.

Conclusion

Drinking fresh Ionized Water, with its micro-clustered molecules, gives the body more anti-oxidant power and deep hydration. Owning a unit also makes it possible to drink as much pure Ionized Water as your body desires every day with no waste of plastic bottles or concern about drinking plastic-contaminated water. And the cost is considerably less over the life of the generator than bottled or designer water – just pennies per gallon.

With its superior hydration, drinking Ionized water promotes good health, oxygenation and detoxification. All-round health and vitality is just a push of a button away, not only for yourself but your entire family. Ionized Water® machines are attractive and modern, simple to install, and come with an outstanding warranty. Simply attach it directly to your kitchen faucet and enjoy cleaner water in the comfort of your own home.

Finally, owning a unit allows you to benefit from the other types of water that can be produced.

Chapter 8

Frequently Asked Questions

Won't my stomach just neutralize Alkaline Water once it reaches the acidic stomach?

This appears to be a logical conclusion. But general physiology suggests that the stomach is not always acidic. Hydrochloric acid is released from parietal cells lining the stomach in response to the presence of both protein and mechanical stretching of the stomach. Water, like alcohol, is absorbed directly across the lining of the stomach without stimulating the release of hydrochloric acid. This suggests that if a person drinks Alkaline Water before meals on an empty stomach, it will be quickly absorbed and circulated throughout the body.

Can I just use baking soda to alkalize my water?

Baking soda is sodium bicarbonate. Bicarbonate is one of the body's major fluid buffers. This seems to be a logical question – in fact, bicarbonate was recently added to one of the newest acid reflux medications as an adjunct to treat acid reflux and neutralize the acidic pH in the stomach. However, electrolyzed, Ionized Water is much more than an alkaline hydrating source. It is rich in antioxidant potential and it is micro-clustered, allowing the water to penetrate cells, much more effectively, introducing both antioxidant and alkaline characteristics to cells desperately in need of relief from oxidative and acidic stress.

How do we begin drinking Ionized water?

It is most common to begin by setting a low pH level (between pH8 and pH9.0), and gradually increasing the amount of intake. After continuing regular intake for 1-2 weeks, gradually adjust the pH level and amount of intake (between pH9.0-pH9.5) based on your bodily needs and physical condition. The ionized water should be as fresh as possible. For children or seniors, set the pH level at 8 and keep it there at least 3-4 weeks before increasing the pH. In the case of infants (until about 2 years old), since breast milk or milk is the only thing they consume, the intestinal movement is generally different from adults, and the use of ionized water is not encouraged until the infant begins to consume more sophisticated foods. In any case, ionized water should not be used to make powdered milk for infants.

What is the white fogginess in ionized water?

The white fogginess apparent in ionized water is simply a result of hydrogen gas being released through electrolysis, and does not pose a problem. If you rest the glass for a while the white fogginess should disappear. Research focused on dissolved hydrogen released through electrolysis of water was presented at The Japan Functional Water Association Meeting in late December of 2002. It reported that ionized water is effective in controlling the oxidation of lipids in the body. It also reported that dissolved hydrogen present in the colloidal state has a higher activity level than active hydrogen produced at the instance of reaction.

The body has multiple buffer systems, why should I expect Ionized Alkaline Water to make a difference?

This is an excellent question that demonstrates sound logic and recognition of the body's major buffering systems. The answer lies in supporting the body's inherent, built-in buffering systems. Remember, the lungs quickly buffer the pH of blood by changing the respiratory rate. The kidneys buffer the blood by excreting hydrogen ions in the urine and absorbing bicarbonate back into the blood. Fluid buffers like bicarbonate and proteins operate much more slowly. Neither the kidneys nor the fluid buffers were designed with the capacity to handle long-term management of the strongly acidic states our bodies are in, owing to poor eating habits, *toxic* environmental exposure, and extreme stress. Alkaline Water closely resembling the pH of some of nature's best foods can help offset many of these challenges.

Should I drink water with a pH of 2.5 to aid stomach digestion, particularly of protein?

No. When drinking water produced by an ionizer, the process of electrolysis by its very chemistry adds electrons to the Alkaline Water and strips them from the acidic water. Therefore, the Acidic Water created from electrolysis is oxidizing. When it is measured with an Oxidation Reduction Potential meter, it will measure as positive (oxidizing). This is the very property that renders Acidic Water bactericidal. Oxidizing products, even water, do not belong in the body as they can directly damage healthy tissue. No one would ever recommend drinking hydrogen peroxide – it has a similar oxidation effect and can injure healthy tissue. Some medical professionals even warn against frequent application of hydrogen peroxide on the skin because it not only destroys bacteria but also healthy tissue, if used in excess. To aid digestion there are some easy things you can do: slow down when you're eating and chew each bite well before swallowing it. Chewing stimulates saliva release and applies mechanical pressure to break the food apart to initiate digestion. Food should be swallowed after it reaches a pasty consistency. For example, if you eat peanuts, chew them to the consistency of peanut butter before swallowing. Eat more of a plant-based diet and try to avoid low-fibre, processed foods that tend to be high in fats, corn syrups and other artificial agents. Eat small meals. Eat only as much as you need. We've been told growing up to finish EVERYTHING that's on our plate. While that is sound advice, it's not very healthy if your plate is overflowing!

Should I drink 11.5 pH water to neutralize my acid reflux?

No, pH 9.5 water would be better. While this approach has the potential to help relieve symptoms, it does not address the underlying problem with physiology. My suggestion is to follow the ideas presented in the section under acid reflux. ALWAYS try to address the root cause of disease and dysfunction. Shortcutting our own physiology is a prescription for disaster!

Can infants drink alkaline water?

The real question is what should infants drink and why? Physiology and nature have answered this question. Infants should ideally nurse from their mothers. The process of milk production, where nutrients are shuttled from the mother to her breast milk to provide the highest quality nutrients for her young, creates the perfect food for infants. In some cases mothers are not able to nurse, necessitating the practice of supplementing with formula. Since the pH of mother's milk ranges from 7.2 to 7.4, it is logical to conclude that ionized water in this pH range is ideal for reconstituting formula because it is free of the contaminants and chlorine found in tap water.

Why do I need to drink pH 7.0 water when I take prescription medications?

Ionized Water that is of pH 7.0 does not support the chemistry that enhances cellular hydration. Medications were designed to be bioavailable to tissues based upon calculations of dissociation constants that are based, not upon ionized water *with* a low NMR or highly negative ORP, but on tap water with a higher NMR and more neutral ORP. Drink only pH 7.0 water with medications and through the next hour to ensure that predicted steady state levels are maintained as indicated by the manufacturer.

Can I take nutritional supplements with ionized water?

Absolutely. In fact, it is believed that the superior hydrating features of the water will deliver the nutrients of the supplement into the cell at a much higher rate and with better efficiency than with tap water. You may be able to reduce the amount of supplements you take.

Is there any danger in drinking too much water?

Yes. There is a condition called Hyponatremia commonly referred to as "Water Intoxication" that results from excessive water consumption. When water intake is excessive, and this can be highly dependent upon individual circumstances, blood sodium levels can become diluted. If sodium levels drop too low, you're likely to experience extreme fatigue, muscle cramps, seizures and swelling in the brain. A 2005 study in the *New England Journal of Medicine* found that close to one sixth of marathon runners develop some degree of *Hyponatremia. This* is a rare medical condition but it can happen with either overconsumption of water or with extreme dehydration. The analogy of changing the water of a fish tank is useful here; if you change too much water in a fish tank at one time. it can shock the fish and kill them. Change your water slowly and consistently. To be safe, do not exceed more than two to three gallons of Ionized Water per day.

Are there any medical conditions that require cautious use of alkaline water?

Yes. Individuals with kidney failure should proceed with caution and consult their physician. However, this group is unquestionably highly acidic and needs the benefits of alkalization, cellular hydration and antioxidant support. Nevertheless, hydrate slowly and with patience.

Individuals with psychosis, who are taking antipsychotic medications, should always be closely monitored by a physician. They are at the biggest risk of water intoxication. Ionized Water is the ideal hydrating source for this group. However, their intake should be monitored closely and they should seek medical supervision to assure they do not develop Hyponatremia. Individuals with Congestive Heart Failure are often on restricted fluid intake and reduced sodium load. If they are closely monitoring their sodium levels, there should be no reason Alkaline Water cannot be the source of water for this group. However, they must consult their physician before implementing Ionized Water into their diet.

How can I get my medical doctor to listen to me when I talk about Ionized Water?

First, respect your doctor's time. Second, if you are granted time, present a well-rehearsed 60 second demonstration illustrating either the ORP, or hydrating properties of the water. Third, and most importantly, doctors want to see scientific evidence to be reassured that what they are being told is not just your personal testimonial, especially if you 're not a peer with a medical degree. That's one of the reasons I wrote this book. I wanted to help my peers and streamline their exposure process by sharing what I've learned. Don't get caught up in the hype of your testimonial stories; it will turn off the critical thinker who is looking for evidence.

Keep the testimonials brief and emphasize the science that supports your testimony.

How can Ionized Water help my athletic performance?

Exercise results in the build-up of metabolic acids. Much of these acids are removed by respiration but muscles in particular build up lactic acid that contributes to fatigue and impairs performance. By drinking water which contains Active Hydrogen, cellular oxidative stress can quickly be reduced, acidic by products like lactic acid can be neutralized and stressed cells can quickly become rehydrated. Water is a necessary component of cellular respiration and nowhere is drinking it more important that for repairing fatigued muscles. *The coach who instructs his team to drink only ionized water is a wise coach and will see better recovery of the athlete both during practice and competition.* I believe ionized water is the best sports nutrition supplement of this century!

Are there any simple ways to explain how, Ionized Water might improve

chronic diseases?

- Oxidative structural damage to collagen crosslinks less skin elasticity and higher wrinkles

- Oxidative structural damage to elastin protein, leading to stiffer

- arterial walls and hypertension

- Oxidative structural damage to arterial wall lining, leading to atherosclerosis

- Oxidative structural damage to DNA/RNA causingmutations,

- leading to inefficient repair processes and cancer

- Oxidative structural damage to Insulin Receptor, leading to diabetes

Ionized Water can prevent oxidative structural damage, thereby protecting you from a whole bunch of chronic illnesses.

What should I look for in a water ionizer?

When selecting an ionizer, it may be helpful to keep the following in mind:

1. A high-quality product will probably not be the least expensive. Be willing to invest in your health.

2. Select a vendor with an established track record. Be cautious of the new kid on the block .

3. Look for industry awards and certifications to ensure quality.

4. A large portion of the cost and quality of an ionizer is found in its interior. Machines with high quality "guts," i. e., platinum-coated titanium plates with a large surface area, will cost more upfront but will consistently produce the healthiest water and last the longest.

5. Most of the less expensive units produce a very slow stream of water.

6. Many of the warranties on inexpensive units are void if the machine is used in areas with hard water (80% of the U.S. has hard water).

7. Many of the warranties on inexpensive units are void if the machine is used commercially.

8. All units will make both alkaline and acidic water. But look for an ionizer that can create strong acidic water with a pH of 2.6 or lower so the bactericidal benefits are available to you.

9. Do your homework on the details such as warranty, return policies and customer support.

10. Be cautious of some of the critiques found on the Internet. They typically lack any credible scientific basis and are written by a biased party with something to gain.

11. Become educated before making your purchase. Attend demonstrations and learn as much as you can first. Talk with others who actually own the model you're thinking of purchasing and get their feedback.

12. Understand that most units will be sold through direct sales marketing. It's not a product you'll find on the shelf of your local big box retailer.

13. Be wary of any vendor that obligates you to sell anything to purchase your unit. Becoming a company distributor should be optional.

14. Ionizers require mineral rich water to generate the many beneficial features we've discussed. If your source water is mineral-free, like reverse osmosis water, your unit will not work properly unless you add back sufficient amount of minerals to the water.

15. Depending on your source water, you may want to purchase pre-filters to extend the life of the filter in the ionizer and deal with other potential contamination issues like fluoride or other contaminants that can exist in well water.

16. Generally speaking, hard water provides better ions with the best ORP values. It is preferred over a soft water supply. But if your water is too hard, it can cause problems for the ionizer

by coating the plates with calcium deposits (like you see in your shower and sink).

17. ORP values decline quickly. It is best to fill a glass of ionized water and drink it immediately. While I have found a container that does help retain ORP values for a couple of days, when stored in plastic bottles the ORP value will typically drop by 50 percent within 24 hours, and revert back to its original state within 48 hours. Micro – clustering features and pH values decline more slowly. depending on the water's exposure to heat and light. By investing in a high – quality machine, you will create higher ORP water to begin with, thus it will retain its ORP values longer. This is very important if you're sharing water with friends and family.

18. With a good ORP meter and pH meter. you can do your own testing to determine how long the various qualities of the water last with your source water and machine.

Chapter 9

Scientific Research Studies on Ionized Alkaline Water

Physiological effects of alkaline ionized water: Effects on metabolites produced by intestinal fermentation

by Takashi Hayakawa, Chicko Tushiya, Hisanori Onoda, Hisayo Ohkouchi, Harul--to Tsuge (Gifu University, Faculty of Engineering, Dept. of Food Science).

We have found that long-term ingestion of alkaline ionized water (AIW) reduces cecal fermentation in rats that were given highly fermentable commercial diet (MF: Oriental Yeast Co., Ltd.). In this experiment, rats were fed MF and test water (tap water, AIW with pH at 9 and 10) for about 3 months. Feces were collected on the 57th day, and the rats were dissected on the 88th day. The amount of ammonium in fresh feces and cecal contents as well as fecal freeglucose tended to drop down for the AIW group. In most cases, the amount of free-amino acids in cecal contents did not differ sign – icantly except for cysteine (decreased in AIW with pH at 10) and isoleucine (increased in AIW with pH at 10).

Purpose of tests

Alkaline ionized water electrolyzers have been approved for manufacturing in 1965 by the Ministry of Health and Welfare of Japan as medical equipment to produce medical substances. Alkaline ionized water (AIW) produced by this equipment is known to be effective

143

against gastrointestinal fermentation, chronic diarrhea, indigestion and hyperchylia as well as for controlling gastric acid.*1 This is mainly based on efficacy of the official calcium hydroxide. *2 By giving AIW to rats for a comparatively long time under the condition of extremely high level of intestinal fermentation, we have demonstrated that AIW intake is effective for inhibition of intestinal fermentation when its level is high based on some test results where AIW worked against cecal hypertrophy and for reduction in the amount of short-chain fatty acid that is the main product of fermentation.*3 We have reported that this is caused by the synergy between calcium level generally contained in AIW (about 50ppm) and the value of pH, and that frequency of detecting some anaerobic bacteria tends to be higher in alkaline ionized water groups than the other, although the bacteria count in the intestine does not have significant difference.

Based on these results, we made a judgment that effect of taking AIW supports part of inhibition mechanism against abnormal intestinal fermentation, which is one of the claims of efficacy that have been attributed to alkaline ionized water electrolyzers. *4 On the other hand, under the dietary condition of low intestinal fermentation, AIW uptake does not seem to inhibit fermentation that leads us to believe that effect of AIW uptake is characteristic of hyperfermentation state. Metabolites produced by intestinal fermentation include indole and skatole in addition to organic acids such as short-chain fatty acid and lactic acid as well as toxic metabolites such as ammonium, phenol and pcresol. We do not know how AIW uptake would affect the production of these materials. In this experiment, we have tested on ammonium production as explained in the following sections.

Testing methods

Four-week-old male Wistar/ST Clean rats were purchased from Japan SLC Co., Ltd. and were divided into 3 groups of 8 each after preliminary breeding. AIW of pH 9 and 10 was produced by an electrolyzer Mineone ROYAL NDX3 1 OH by Omco Co., Ltd. This model produces AIW by electrolyzing water with calcium lactate added. On the last day of testing, the rats were dissected under Nembutal anesthesia to take blood from the heart by a heparin-treated syringe. As to their organs, the small intestines, cecum and colon plus rectum were taken out from each of them. The cecurn was weighed and cleaned with physiological saline after its contents were removed, and the tissue weight was measured after wiping out moisture. Part of cecal contents was measured its pH, and the rest was used to assay ammonium concentration. The amount of ammonium contained in fresh feces and cecal contents was measured by the Nessler method after collecting it in the extracted samples using Conway 's micro-diffusion container. Fecal freeglucose was assayed by the oxygen method after extraction by hot water. Analysis of free amino acids contained in cecal contents was conducted by the Waters PicoTag amino acid analysis system.

Test results and analyses

No difference was found in the rats' weight gain, water and feed intake and feeding efficiency, nor was any particular distinction in appearance identified. The length of the small intestines and colon plus rectum tended to decline in AIW groups. PH value of cecal contents was higher and the amount of fecal free-glucose tended to be lower in AIW groups than the control group. Since there was no difference in fecal discharge itself, the amount of free-glucose discharged per day was at a low level. The amount of discharged free-glucose in feces is greater when intestinal fermentation is more intensive, which indicates that intestinal fermentation is more inhibited in AIW groups than the control group. Ammonium concentration in cecal contents tends to drop down in AIW groups (Fig. 1). This trend was most distinctive in case of fresh feces of one of AIW groups with pH 10 (Fig.2) AIW uptake was found to be inhibitory against ammonium production. In order to study dynamics of amino acids in large intestines, we examined free amino acids in the cecal contents to find out that cysteine level is low in AIW groups whereas isoleucine level is high in one of AIW groups with pH 10, although no significant difference was identified for other amino acids.

Bibliography

1. "Verification of Alkaline Ionized Water" by Life Water Institute, Metamor Publishing Co., 1994, p.46 2. "Official Pharmaceutical Guidelines of Japan, Vol. IT' by Japan Public Documents Association, Hirokawa PublIshin Co. , 1996 3. "Science and Technology of Functional Water" (part) by Takashi Hayakawa, Haruffito Tsuge, edited by Water Scienll cc Institute, 1999, pp.109-116 4. "Tasics and Effective Use of Alkaline Ionized Water" by Takashi Hayakawa, Haruhito Tsuge, edited by Tetsuji Hc kudou, 25th General Assembly of Japan Medical Congress 'Tunctional Water in Medical Treatment", Administratio~ Offices, 1999, pp. 10 – 11.

Adequate fluid replacement helps maintain hydration and, promotes the health, safety, and optimal physical performance of individuals participating in regular physical activity Med Sci Sports Exercise 1996 Jan;28(1):i-vii.

American College of Sports Medicine position stand. Exercise and fluid replacement.

Convertino VA, Armstrong LE, Coyle EF, Mack GW, Sawka MN, Senay LC Jr, Sherman WM.

It is the position of the American College of Sports Medicine that adequate fluid replacement helps maintain hydration and, therefore, promotes the health, safety, and optimal physical performance of individuals participating in regular physical activity. This position statement is based on a comprehensive review and interpretation of scientific literature concerning the influence of fluid replacement on exercise performance and the risk of thermal injury associated with dehydration and hyperthermia.

Based on available evidence, the American College of Sports Medicine makes the following general recommendations on the amount and composition of fluid that should be ingested in preparation for, during, and after exercise or athletic competition:

1. It is recommended that individuals consume a nutritionally balanced diet and drink adequate fluids during the 24-hr period before an event, especially during the period that includes the meal prior to exercise, to promote proper hydration before exercise or competition.

2. It is recommended that individuals drink about 500 ml (about 17 ounces) of fluid about 2 h before exercise to promote adequate hydration and allow time for excretion of excess ingested water.

3. During exercise, athletes should start drinking early and at regular intervals in an attempt to consume fluids at a rate sufficient to replace all the water lost through sweating (i.e., body weight loss), or consume the maximal amount that can be tolerated.

4. It is recommended that ingested fluids be cooler than ambient temperature [between 15 degrees and 22 degrees C (59 degrees and 72 degrees F])] and flavored to enhance palatability and promote fluid

5. replacement. Fluids should be readily available and served in containers that allow adequate volumes to be ingested with ease and with minimal interruption of exercise.

6. Addition of proper amounts of carbohydrates and/or electrolytes to a fluid replacement solution is recommended for exercise events of duration greater than 1 h since it does not significantly impair water delivery to the body and may enhance performance. During exercise lasting less than 1 h, there is little evidence of physiological or physical performance differences between consuming a carbohydrate-electrolyte drink and plain water.

7. During intense exercise lasting longer than 1 h, it is recommended that carbohydrates be ingested at a rate of 30-60 g.h(-1) to maintain oxidation of carbohydrates and delay fatigue. This rate of carbohydrate intake can be achieved without compromising fluid delivery by drinking 600 – 1200 ml.h(-1) of solutions containing 4%-8% carbohydrates (g.100 ml(-1)). The carbohydrates can be sugars (glucose or sucrose) or starch (e.g., maltodextrin).

8. Inclusion of sodium (0.5-0.7 g.1(-1) of water) in the rehydration solution ingested during exercise lasting longer than 1 h is recommended since it may be advantageous in enhancing palatability, promoting fluid retention, and possibly preventing hyponatremia in certain individuals who drink excessive quantities of fluid. There is little physiological basis for the presence of sodium in n oral rehydration solution for enhancing intestinal water absorption as long as sodium is sufficiently available from the previous meal.

Electrolyzed-reduced water scavenges active oxygen species and protects DNA from oxidative damage

Biochem Biophys Res Commun. , 1997 May 8;234(1):269-74.

Shirahata S , Kabayama S, Nakano M, Miura T, Kusumoto K, Gotoh M, Hayashi H , Otsubo K, Morisawa S, Katakura Y.

Institute of Cellular Regulation Technology, Graduate School of Genetic Resources Technology, Kyushu University, Fukuoka, Japan . sirahata@grt. kyushu-u.ac.jp

Active oxygen species or free radicals are considered to cause extensive oxidative damage to biological macromolecules, which brings about a variety of diseases as well as aging. The ideal scavenger for active oxygen should be 'active hydrogen'. 'Active hydrogen' can be produced in reduced water near the cathode during electrolysis of water. Reduced water exhibits high pH, low dissolved oxygen (DO), extremely high dissolved molecular hydrogen (DH), and extremely negative redox potential (RP) values. Strongly electrolyzed-reduced water, as well as ascorbic acid, (+)-catechin and tannic acid, completely scavenged O.-2 produced by the hypoxanthinexanthine oxidase (HX-XOD) system in sodium phosphate buffer (pH 7.0).

The superoxide dismutase (SOD)-like activity of reduced water is stable at 4 degrees C for over a month and was not lost even after neutralization, repeated freezing and melting, deflation with sonication, vigorous mixing, boiling, repeated filtration, or closed autoclaving, but was lost by opened autoclaving or by closed autoclaving in the presence of tungsten trioxide which efficiently adsorbs active atomic hydrogen. Water bubbled with hydrogen gas exhibited low DO, extremely high DH and extremely low RP values, as does reduced water, but it has no SOD-like activity. These results suggest that the SOD-like activity of reduced water is not due to the dissolved molecular hydrogen but due to the dissolved atomic hydrogen (active hydrogen). Although SOD accumulated H2O2 when added to the HX-XOD system, reduced water decreased the amount of H2O2 produced by XOD. Reduced water, as well as catalase and ascorbic acid, could directly scavenge H2O2. Reduced water suppresses single-strand breakage of DNA b active oxygen species produced by the Cu(II)-catalyzed oxidation of ascorbic acid in a dose-dependent manner, suggesting that reduced water can scavenge not only O2. – and H2O2, but also 1O2 and .OH.

PMID: 9169001 [PubMed – indexed for MEDLINE]

The mechanism of the enhanced antioxidant effects against superoxide anion radicals of reduced water produced by electrolysis

Biophys Chem. 2004 Jan 1;107(1):71-82.

Hanaoka K, Sun D, Lawrence R, Kamitani Y, Fernandes G.

Bio-REDOX Laboratory Inc. 1187-4, Oaza-Ueda, Ueda-shi, Nagano-ken 386 – 0001, Japan . hanak@rapid.ocn.ne.jp

We reported that reduced water produced by electrolysis enhanced the antioxidant effects of proton donors such as ascorbic acid (AsA) in a previous paper. We also demonstrated that reduced water produced by electrolysis of 2 mM NaCl solutions did not show antioxidant effects by itself. We reasoned that the enhancement of antioxidant effects may be due to the increase of the ionic product of water as solvent. The ionic product of water (pKw) was estimated by measurements of pH and by a neutralization titration method. As an indicator of oxidative damage, Reactive Oxygen Species – (ROS) mediated DNA strand breaks were measured by the conversion of supercoiled phiX-174 RF I double-strand DNA to open and linear forms. Reduced water had a tendency to suppress single-strand breakage of DNA induced by reactive oxygen species produced by H2O2/Cu (II) and HQ/Cu (II) systems. The enhancement of superoxide anion radical dismutation activity can be explained by changes in the ionic product of water in the reduced water.

PMID: 14871602 [PubMed – in process]

Comparison of electrolyzed oxidizing water with various antimicrobial interventions to reduce Salmonella species on poultry

Poult Science 2002 Oct;81(10):1598-605.

Fabrizio KA, Sharma RR, Demirci A, Cutter CN.

Department of Food Science, The Pennsylvania State University, University Park 16802 , USA .

Foodborne pathogens in cell suspensions or attached to surfaces can be reduced by electrolyzed oxidizing (EO) water; however, the use of EO water against pathogens associated with poultry has not been explored. In this study, acidic EO water [EO-A; pH 2.6, chlorine (CL) 20 to 50 ppm, and oxidation-reduction potential (ORP) of 1,150 mV], basic EO water (EO-B; pH 11.6, ORP of – 795 mV), CL, ozonated water (OZ), acetic acid (AA), or trisodium phosphate (TSP) was applied to broiler carcasses inoculated with Salmonella Typhimurium (ST) and submerged (4 C, 45 min), spray-washed (85 psi, 25 C, 15 s), or subjected to multiple interventions (EO-B spray, immersed in EO-A; AA or TSP spray, immersed in CL). Remaining bacterial populations were determined and compared at Day 0 and 7 of aerobic, refrigerated storage. At Day 0, submersion in TSP and AA reduced ST 1.41 log10, whereas EO-A water reduced ST approximately 0.86 log10. After 7 d of storage, EO-A water, OZ, TSP, and AA reduced ST, with detection only after selective enrichment.

Spray-washing treatments with any of the compounds did not reduce ST at Day 0. After 7 d of storage, TSP, AA, and EO-A water reduced ST 2.17, 2.31, and 1.06 log10, respectively. ST was reduced 2.11 log10 immediately following the multiple interventions, 3.81 log10 after 7 d of storage. Although effective against ST, TSP and AA are costly and adversely affect the environment. This study demonstrates that EO water can reduce ST on poultry surfaces following extended refrigerated storage.

PMID: 12412930 [PubMed – indexed for MEDLINE]

Inactivation of Escherichia coli (O157:H7) and Listeria monocytogenes on plastic kitchen cutting boards by electrolyzed oxidizing water

Venkitanarayanan KS , Ezeike GO, Hung YC, Doyle MP.

Department of Animal Science, University of Connecticut, Storrs 06269 , USA .

One milliliter of culture containing a five-strain mixture of Escherichia coli O157:H7 (approximately 10(10) CFU) was inoculated on a 100-cm2 area marked on unscarred cutting boards. Following inoculation, the boards were air-dried under a laminar flow hood for 1 h, immersed in 2 liters of electrolyzed oxidizing water or sterile deionized water at 23 degrees C or 35 degrees C for 10 or 20 min; 45 degrees C for 5 or 10 min; or 55 degrees C for 5 min. After each temperature-time combination, the surviving population of the pathogen on cutting boards and in soaking water was determined.

Soaking of inoculated cutting boards in electrolyzed oxidizing water reduced E. coli O157:H7 populations by > or = 5.0 log CFU/100 cm2 on cutting boards. However, immersion of cutting boards in deionized water decreased the pathogen count only by 1.0 to 1.5 log CFU/100 cm2. Treatment of cutting boards inoculated with Listeria monocytogenes in electrolyzed oxidizing water at selected temperature-time combinations (23 degrees C for 20 min, 35 degrees C for 10 min, and 45 degrees C for 10 min) substantially reduced the populations of L. monocytogenes in comparison to the counts recovered from the boards immersed in deionized water. E. coli O157:H7 and L. monocytogenes were not detected in electrolyzed oxidizing water after soaking treatment, whereas the pathogens survived in the deionized water used for soaking the cutting boards. This study revealed that immersion of kitchen cutting boards in electrolyzed oxidizing water could be used as an effective method for inactivating foodborne pathogens on smooth, plastic cutting boards.

PMID: 10456736 [PubMed – indexed for MEDLINE]

The bactericidal effects of electrolyzed oxidizing water on bacterial strains involved in hospital infections

Vorobjeva NV , Vorobjeva LI, Khodjaev EY.

Artificial Organs 2004 Jun;28(6):590-2.

Department of Physiology of Microorganisms, Biology Faculty, Moscow State University, Lenin Hills 1/12, Moscow 119992, Russia. nvvorobjeva@ mail.ru

The study is designed to investigate bactericidal actions of electrolyzed oxidizing water on hospital infections. Ten of the most common opportunistic pathogens are used for this study. Cultures are inoculated in 4.5 mL of electrolyzed oxidizing (EO) water or 4.5 mL of sterile deionized water (control), and incubated for 0, 0.5, and 5 min at room temperature. At the exposure time of 30 s the EO water completely inactivates all of the bacterial strains, with the exception of vegetative cells and spores of bacilli which need 5 min to be killed. The results indicate that electrolyzed oxidizing water may be a useful disinfectant for hospital infections, but its clinical application has still to be evaluated. PMID: 15153153 [PubMed – in process]

Effect of electrolyzed oxidizing water and hydrocolloid occlusive dressings on excised burn-wounds in rats

Chin J Traumatol . 2003 Aug 1;6(4):234-7.

Xin H, Zheng YJ, Hajime N, Han ZG.

Department of Thoracic Surgery, China – Japan Union Hospital, Jilin University, Jilin 130031, China. xinhua7254@yahoo.com.cn

OBJECTIVE: To study the efficacy of electrolyzed oxidizing water (EOW) and hydrocolloid occlusive dressings in the acceleration of epithelialization in excised burn-wounds in rats.

METHODS: Each of the anesthetized Sprague-Dawley rats (n=28) was subjected to a thirddegree burn that covered approximately 10% of the total body surface area. Rats were assigned into four groups: Group I (no irrigation), Group II (irrigation with physiologic saline), Group III (irrigation with EOW) and Group IV (hydrocolloid occlusive dressing after EOW irrigation).

Wounds were observed macroscopically until complete epithelialization was present, then the epithelialized wounds were examined microscopically.

RESULTS: Healing of the burn wounds was the fastest in Group IV treated with hydrocolloid occlusive dressing together with EOW. Although extensive regenerative epidermis was seen in each Group, the proliferations of lymphocytes and macrophages associated with dense collagen deposition were more extensive in Group II, III and IV than in Group I. These findings were particularly evident in Group III and IV.

CONCLUSIONS: Wound Healing may be accelerated by applying a hydrocolloid occlusive dressing on burn surfaces after they are cleaned with EOW.

PMID: 12857518 [PubMed – indexed for MEDLINE]

Use of Ionized water in hypochlorhydria or achlorhydria

Prof. Kuninaka Hironage, Head of Kuninaka Hospital

"Too many fats in the diets, which lead to the deposition of cholesterol on the blood vessels, which in turn constrict the blood flow, cause most illnesses such as high blood pressure. In accordance with the theory of Professor Gato of Kyushu University on Vitamin K (because vitamin K enables the blood calcium to increase) , or the consumption of more antioxidant water, the effectiveness of the increase in the calcium in high

blood pressure is most significant. The consumption of alkaline antioxidant water for a period of 2 to 3 months, I have observed the blood pressure slowly drop, due to the water's solvent ability, which dissolves the cholesterol in the blood vessels."

Effect of electrolyzed water on wound healing

Artif Organs. 2000 Dec;24(12):984-7.

Yahagi N, Kono M, Kitahara M, Ohmura A, Sumita O, Hashimoto T, Hori K, Ning-Juan C, Woodson P, Kubota S, Murakami A, Takamoto S.

Department of Anesthesiology, Teikyo University Mizonokuchi Hospital, Tokyo, Japan . naokiyah@aol.com Electrolyzed water accelerated the healing of full-thickness cutaneous wounds in rats, but only anode chamber water (acid pH or neutralized) was effective. Hypochlorous acid (HOCl), also produced by electrolysis, was ineffective, suggesting that these types of electrolyzed water enhance wound healing by a mechanism unrelated to the well-known antibacterial action of HOCl. One possibility is that reactive oxygen species, shown to be electron spin resonance spectra present in anode chamber water, might trigger early wound healing through fibroblast migration and proliferation.

PMID: 11121980 [PubMed – indexed for MEDLINE]

Allergies and Ionized Water

Prof. Kuninaka Hironaga, Head of Kuninaka Hospital

"Mr. Yamada, the head of Police Research Institute, suffered from severe allergy. He was treated repeatedly by skin specialist, but with no success. Then he started consuming antioxidant water. The allergy responded very well and was soon completely cured. No relapse had occurred, although he had taken all kinds of food. He was most grateful and excited about this treatment.

As for myself, I had also suffered severe allergy. Ever since I began to consume antioxidant water, the allergy has recovered. Since then, I started a research on the effectiveness of antioxidant water.

I discovered that most allergies are due to acidification of body condition and is also related to consuming too much meat and sugar. In every allergy case, the patient's antioxidant minerals are excessively low which in turn lower the body resistance significantly. The body becomes overly sensitive and develops allergy easily. To stabilize the sensitivity, calcium solution in injected into the vein. Therefore, it is clear that the antioxidant water has ionic calcium, which can help alleviate allergy.

The ionic calcium not only enhances the heart, urination, and neutralization of toxins but controls acidity. It also enhances the digestive system and liver function. This will promote natural healing power and hence increase its resistance to allergy. In some special cases of illness, which do not respond to drugs, it is found, it is found to respond well to antioxidant water."

Digestive Problems and Ionized Water

Prof. Kogure Keizou, Kogure Clinic of Juntendo Hospital

"The stomach is readily upset both by diseases affecting the stomach and by other general illnesses. In addition, any nervous tension or anxiety frequently causes gastric upset, vague symptoms when This information is under some strain.

The important role of antioxidant water in our stomach is to neutralize the secretion and strengthen it s functions. Usually, after consuming the antioxidant water for 1 to 3 minutes, the gastric juice increase to 1½ times.

For those suffering from hypochlorhydria or achlorhydria (low in gastric juice) the presence of antioxidant water will stimulate the stomach cells to secrete more gastric juice. This in turn enhances digestion and absorption of minerals.

However, on the other hand, those with hyperchlorhydria (high in gastric juice), the antioxidant water neutralizes the excessive gastric juice. Hence, it does not create any adverse reaction.

According to the medical lecturer from Maeba University, the pH of the gastric secretion will still remain normal when antioxidant water is consumed. This proves that the ability of the antioxidant water is able to neutralize as well as to stimulate the secretion."

Use of Ionized water for gynecological conditions and treatments

Prof. Watanabe Ifao, Watanabe Hospital

"Ionized alklaine antioxidant water improves body constituents and ensures effective healing to many illnesses. The uses of antioxidant water in gynecological patients have proved to be very effective. The main reason for its effectiveness is that this water can neutralize toxins.

When given antioxidant water to pre-eclamptic toxemia cases, the results are most significant. During my long years of servicing the pre-eclamptic toxemia cases, I found that the women with pre-eclamptic toxemia who consumed antioxidant water tend to deliver healthier babies with stronger muscles. A survey report carried out on babies in this group showed intelligence above average."

Toxin Neutralization with the Use of Electrolysed Water

Prof. Kuwata Keijiroo, Doctor of Medicine

"In my opinion, the wonder of antioxidant water is the ability neutralizes toxins, but it is not a medicine. The difference is that the medicine can only apply to each and individual case, whereas the antioxidant water can be consumed generally and its neutralizing power is something which is very much unexpected. Now, in brief, let me introduce to you a heart disease case and how it was cured.

The patient was a 35 years old male suffering from vascular heart disease. For 5 years, his sickness deteriorated. He was in the Setagays Government Hospital for treatment.

During those 5 years, he had been in and out of the hospital 5 to 6 times. He had undergone high tech examinations such as angiogram by injecting VINYL via the vein into the heart. He consulted and sought treatment from many good doctors where later he underwent a major surgical operation. Upon his discharge from the hospital, he quit his job to convalesce.

However, each time when his illness relapsed, the attack seemed to be even more severe.

Last year, in August, his relatives were in despair and expected he would not live much longer. It so happened at that time that the victim's relative came across antioxidant water processor. His illness responded well and he is now on the road to recovery."

In the United States, cardiovascular diseases account for more than one-half of the approximate 2 million deaths occurring each year.... It is estimated that optimal conditioning of drinking water could reduce this cardiovascular disease mortality rate by as much as 15 percent in the United States Report of the Safe Drinking Water Committee of the National Academy of Sciences, 1977

Eczema and the Effects of Ionized Water

Prof. Tamura Tatsuji, Keifuku Rehabilitation Center "Eczema is used to describe several varieties of skin conditions, which have a number of common features. The exact cause or causes of eczema are not fully understood. I many cases, eczema can be attributed by external irritants.

Let me introduce a patient who recovered from skin disease after consuming the antioxidant water. This patient suffered 10 years of eczema and could not be cured effectively even under specialist treatment. This patient, who is 70 years of age, is the president of a vehicle spare parts company. After the war, his lower limbs suffered acute eczema, which later became chronic. He was repeatedly treated in a specialist skin hospital.

The left limb responded well to treatment, but not so on the right limb. He suffered severe itchiness, which, when scratched led to bleeding. During the last 10 years, he was seen and treated by many doctors. When I first examined him, his lower limb around the joints was covered with vesicles. Weeping occurred owing to serum exuding from the vesicles.

I advised him to try consuming antioxidant water. He bought a unit and consumed the antioxidant water religiously and used the acidic water to bathe the affected areas. After 2 weeks of treatment the vesicles dried up. The eczema was completely cleared without any relapse after 1½ month."

Diabetes and the Effects of Ionized Water

Prof. Kuwata Keijiroo, Doctor of Medicine

"When I was serving in the Fire Insurance Association, I used to examine many diabetic patients. Besides treating them with drugs, I provided them with antioxidant water. After drinking antioxidant water for one month, 15 diabetic patients were selected and sent to Tokyo University for further test and observations.

Initially, the more serious patients were a bit apprehensive about the treatment. When the antioxidant water was consumed for some time, the sugar in the blood and urine ranged from a ratio of 300 mg/l to 2 mg/dc. There was a time where the patient had undergone 5 to 6 blood tests a day and detected to be within normal range. Results also showed that even 1 ½ hour after meals, the blood sugar and urine ratio was 100 mg/dc: 0 mg/dc. The sugar in the urine has completely disappeared."

NOTE: More Americans than ever before are suffering from diabetes, with the number of new cases averaging almost 800,000 each year. The disease has steadily increased in the United States since 1980, and in 1998, 16 million Americans were diagnosed with diabetes (10.3 million diagnosed; 5.4 million undiagnosed). Diabetes is the seventh leading cause of death in the United States, and more than 193,000 died from the disease and its related complication in 1996. ....

The greatest increase – 76 percent – occurred in people age 30 to 30. ....

From: U. S. Department of Health and Human Services, October 13, 2000 Fact Sheet.

Use of Ionized water in treating Acidosis

Prof. Hatori Tasutaroo, Head of Akajiuiji Blood Centre, Yokohama Hospital, Faitama District

" Due to a higher standard of living, our eating habits have changed.

We consume too much proteins, fats and sugar. The excess fats and carbohydrates are in the body as fats. In the present lifestyles, Americans are more extravagant on food compared to the Japanese. Due to this excessive intake obesity is a significant problem. Normally, one out of five males and one out of four females is obese.

The degree of "burn-out" in food intake largely depends on the amount on intake of vitamins and minerals. When excessive intake of proteins, carbohydrates and fats occurs, the requirement for vitamins and minerals increases. However, there is not much research carried out pertaining to the importance of vitamins and minerals.

Nowadays, many people suffer from acidification that leads to diabetes, heart diseases, cancer, live and kidney diseases. If our food intake can be completely burned off, then there is no deposition of fats. Obviously, there will be no acidification problem and hence there should not be any sign of obesity.

The antioxidant water contains an abundance of ionic calcium. This ionic calcium helps in the "burn-off" process. By drinking antioxidant water, it provides sufficient minerals for our body. As a result, we do not need to watch our diet to stay slim.

Hence, antioxidant water is a savior for those suffering from obesity and many adult diseases, providing good assistance in enhancing good health."

Reduced Water for the Prevention of Disease

Dr. Sanetaka Shirahata Graduate school of Genetic Resources Technology, Kyushu University, 6-10-1 Hakozaki, Higashi-ku, Fukuoka 812-8581, Japan.

It has long been established that reactive oxygen species (ROS) cause many types of damage to biomolecules and cellular structures that, in turn result in the development of a variety of pathologic states such as diabetes, cancer and aging. Reduced water is defined as antioxidative water produced by reduction of water. Electrolyzed reduced water (ERW) has been demonstrated to be hydrogen-rich water and can scavenge ROS in vitro (Shirahata et al., 1997).

The reduction of proton in water to active hydrogen (atomic hydrogen, hydrogen radical) that can scavenge ROS is very easily caused by a weak current, compared to oxidation of hydroxyl ion to oxygen molecule.

Activation of water by magnetic field, collision, minerals etc. will also produce reduced water containing active hydrogen and/or hydrogen molecule. Several natural waters such as Hita Tenryosui water drawn from deep underground in Hita city in Japan, Nordenau water in Germany and Tlacote water in Mexico are known to alleviate various diseases.

We have developed a sensitive method by which we can detect active hydrogen existing in reduced water, and have demonstrated that not only ERW but also natural reduced waters described above contain active hydrogen and scavenge ROS in cultured cells. ROS is known to cause reduction of glucose uptake by inhibiting the insulin-signaling pathway in cultured cells. Reduced water scavenged intracellular ROS and stimulated glucose uptake in the presence or absence of insulin in both rat L6 skeletal muscle cells and mouse 3T3/L1 adipocytes. This insulin-like activity of reduced water was inhibited by wortmannin that is specific inhibitor of PI-3 kinase, a key molecule in insulin signaling pathways. Reduced water protected insulinresponsive cells from sugar toxicity and improved the damaged sugar tolerance of type 2 diabetes model mice, suggesting that reduced water may improve insulin-independent diabetes mellitus.

Cancer cells are generally exposed to high oxidative stress. Reduced water cause impaired tumor phenotypes of human cancer cells, such as reduced growth rate, morphological changes, reduced colony formation ability in soft agar, passage number-dependent telomere shortening, reduced binding abilities of telomere binding proteins and suppressed metastasis.

Reduced water suppressed the growth of cancer cells transplanted into mice, demonstrating their anti-cancer effects in vivo. Reduced water will be applicable to not only medicine but also food industries, agriculture, and manufacturing industries.

Shirahata, S. et al .: Electrolyzed reduced water scavenges active oxygen species and protects DNA from oxidative damage. Biochem. Biophys. Res. Commun., 234, 269174, 1997.

Clinical Impovements Obtained From The Intake Of Reduced Water

Extracts from "Presentation At The Eight Annual International Symposium On man And His Environment in Health And Disease" on February 24th 1990, at The Grand Kempinski Hotel, Dalls, Texas, USA by Dr. H. Hayashi, M.D. and Dr. M Kawamura, M.D., on:-

Since the introduction of alkaline ionic water in our clinic in 1985, we have had the following interesting clinical experiences in the use of this type of water. By the use of alkaline ionic water for drinking and the preparation of meals for our in-patients, we have noticed:

- Declines in blood sugar levels in diabetic patients.

- Improvements in peripheral circulation in diabetic gangrene. Declines in uric acid levels in patients with gout.

- Improvements in liver function exams in hepatic disorders.

- Improvements in gastroduodenal ulcer and prevention of their recurrences.

- Improvements in hypertension and hypotension.

- Improvements in allergic disorders such as asthma, urticaria, rhinites and atopic dermatitis.

- Improvements in persistent diarrhoea which occurred after gastrectomy.

- Quicker improvements in post operative bower paralysis.

- Improvements in serum bilirubin levels in new born babies.

Being confirming clinical improvements, we have always observed changes of stools of the patients, with the colour of their feaces changing from black-brown colour to a brigher yellowbrown one, and the odour of their feaces becoming almost negligible.

The number of patients complaining of constipation also decreased markedly. The change of stool findings strongly suggests that alkaline ionic water intake can decrease the production of putrefield or pathogenic metabolites.

Devices to produce reduced water were introduced into our clinic in May 1985. Based on the clinical experiences obtained in the past 15 years, it can be said that introduction of electrolyzed-reduced water for drinking and cooking purpose for in-patients should be the very prerequisite in our daily medical practices. Any dietary recipe cannot be a scientific one if property of water is not taken by the patients is not taken into consideration.

The Ministry of Health and Welfare in Japan announced in 1965 that the intake of reduced water is effective for restoration of intestinal flora metabolism.

Clinical evaluation of alkaline ionized water for abdominal complaints: Placebo controlled double blind tests

by Hirokazu Tashiro, Tetsuji Hokudo, Hiromi Ono, Yoshihide Fujiyama, Tadao Baba (National Ohkura Hospital, Dept. of Gastroenterology; Institute of Clinical Research, Shiga University of Medical Science, Second Dept. of Internal Medicine)

Effect of alkaline ionized water on abdominal complaints was evaluated by placebo controlled double blind tests. Overall scores of improvement using alkaline ionized water marked higher than those of placebo controlled group, and its effect proved to be significantly higher especially in slight symptoms of chronic diarrhea and abdominal complaints in cases of general malaise. Alkaline ionized water group did not get interrupted in the course of the test, nor did it show serious side effects, nor abnormal test data. It was confirmed that alkaline ionized water is safer and more effective than placebos.

Summary

Effect of alkaline ionized water on abdominal complaints was clinically examined by double blind tests using clean water as placebo. Overall improvement rate was higher for alkaline ionized water group than placebo group and the former proved to be significantly more effective than the other especially in cases of slight symptoms. Examining improvement rate for each case of chronic diarrhea, constipation and abdominal complaints, alkaline ionized water group turned out to be more effective than placebo group for chronic diarrhea, and abdominal complaints. The test was stopped in one case of chronic diarrhea, among placebo group due to exacerbation, whereas alkaline ionized water group did not stop testing without serious side effects or abnormal test data in all cases. It was confirmed that alkaline ionized water is more effective than clean water against chronic diarrhea, abdominal complaints and overall improvement rate (relief of abdominal complaints) and safer than clean water.

Introduction

Since the approval of alkaline ionized water electrolyzers by Pharmaceutical Affairs Law in 1966 for its antacid effect and efficacy against gastrointestinal disorders including hyperchylia, indigestion, abnormal gastrointestinal fermentation and chronic diarrhea, they have been extensively used among patients. However, medical and scientific evaluation of their validity is not established. In our study, we examined clinical effect of alkaline ionized water on gastrointestinal disorders across many symptoms in various facilities. Particularly, we studied safety and usefulness of alkaline ionized water by double blind tests using clean water as a control group.

Test subjects and methods

163 patients (34 men, 129 women, age 21 to 72, average 38.6 years old) of indigestion, abnormal gastrointestinal fermentation (with abnormal gas emission and rugitus) and abdominal complaints caused by irregular dejection (chronic diarrhea, or constipation) were tested as subjects with good informed consent. Placebo controlled double blind tests were conducted using alkaline ionized water and clean water at multiple facilities. An alkaline ionized water electrolyzer sold commercially was installed with a pump driven calcium dispenser in

each of the subject homes. Tested alkaline ionized water had pH at 9.5 and calcium concentration at 30ppm. Each subject in placebo group used a water purifier that has the same appearance as the electrolyzer and produces clean water.

The tested equipment was randomly assigned by a controller who scaled off the key code which was stored safely until the tests were completed and the seal was opened again.

Water samples were given to each patient in the amount of 200ml in the morning with the total of 500ml or more per day for a month. Before and after the tests, blood, urine and stool were tested and a log was kept on the subjective symptoms, bowel movements and accessory symptoms.

After the tests, the results were analyzed based on the log and the test data.

Test Results

Symptom

Among 163 tested subjects, alkaline ionized water group included 84 and placebo group 79. Background factors such as gender, age and basal disorders did not contribute to significant difference in the results.

Overall improvement rate

As to overall improvement rate of abdominal complaints, alkaline ionized water group had 2 cases of outstanding improvement (2.5%), 26 cases of fair improvement (32.1%), 36 cases of slight improvement (44.4%), 13 cases of no change (16%) and 4 cases of exacerbation (4.9%), whereas placebo group exhibited 4 (5.2%), 19 (24.7%), 27 (35.1%), 25 (32.5%) and 2 cases (2.6%) for the same category. Comparison between alkaline ionized water and placebo groups did not reveal any significant difference at the level of 5% significance according to the Wilcoxon test, although alkaline ionized water group turned out to be significantly more effective than placebo group at the level of p value of 0.22.

Examining overall improvement rates by a 7, 2 test (with no adjustment for continuity) between the effective and noneffective groups, alkaline ionized water group had 64 (79%) of effective cases and 17 cases (21%) of non effective cases, whereas placebo group had 50 (64.9%) and 27 (35.1%) cases respectively. The result indicated that alkaline ionized water group was significantly more effective than placebo group at the level of p value of 0.0.48.

Looking only at 83 slight cases of abdominal complaints, overall improvement rate for alkaline ionized water group (45 cases) was composed of 11 cases (242%) of fair improvement, 22 cases (48.9%) of slight improvement, 17 cases (44.7%) of no change and 3 cases (6.7%) of exacerbation, whereas placebo group (38 cases) had 3 (7.8%), 17 (44.7%), 17 (44.7%) and 1 (2.6%) cases for the same category. Alkaline ionized water group was significantly more effective than placebo group according to the comparison between the groups (p value = 0.033).

Improvement rate by basal symptom

Basal symptoms were divided into chronic diarrhea, constipation and abdominal complaints (dyspepsia) and overall improvement rate was evaluated for each of them to study effect of alkaline ionized water. In case of chronic diarrhea, alkaline ionized water group resulted in 94.1% of effective cases and 5.9% of non effective cases. Placebo group came up with 64,7% effective and 35.3% non effective. These results indicate alkaline ionized water group proved to be significantly more effective than placebo group. In case of slighter chronic diarrhea, comparison between groups.